Diesel-Electric Locomotives

Diesel-Electric Locomotives

How They Work, Use Energy, and Can Become More
Efficient and Environmentally Sustainable

by Walter Simpson

Simmons-Boardman Books, Inc.
Omaha, NE

ISBN 9780911382693

Library of Congress Control Number: 2018963271

Cover and book design by Robert Hochberg.

Visit the accompanying website for this book: www.diesel-electric-locomotives.com.

Published and distributed by:
Simmons-Boardman Books, Inc.
The Railway Educational Bureau
1809 Capitol Avenue
Omaha, NE 68102
402-346-4300 or 800 228-9670
Email: orders@transalert.com
Website: www.transalert.com

Printed in the United States of America
First printing 2018.

CONTENTS
AT A GLANCE

Detailed Table of Contents

Diesel-Electric Locomotives
How They Work, Use Energy, and Can Become More Efficient and Environmentally Sustainable

Thurmond, WV, June 2014. Photo by author.

Foreword

Passionate people are hard to come by. For this reason, I often find myself immediately drawn to people who live life with passion. Their lives reflect both a rigorous pursuit of knowledge and the tangible and visible application of that knowledge. Walter Simpson is a person of passion.

I first met Walter in 2014 when he and his wife Nan made the 4-hour drive south from Amherst, NY, in their Toyota Prius to meet with my colleague and me in Altoona, PA. After only a few short minutes of conversation, I learned about two of Walter's many passions: Energy and trains. I also discovered that he and Nan not only drive a hybrid vehicle, but they live in a 100% solar-powered home! After a remarkable career in the field of university energy conservation and environmental sustainability, Walter has turned his attention to locomotives.

When Walter told me that he was writing a book on diesel-electric locomotives, I was immediately interested. I teach at Penn State Altoona in the nation's first and only accredited baccalaureate program in Rail Transportation Engineering (RTE). As an educator, I'm always looking for textbooks and resources that will enhance the learning experience of our students. Unlike traditional Civil, Mechanical, or Electrical Engineering programs, the RTE curriculum includes specialized railroad engineering courses focused on track, signals, and rolling stock. We are challenged to find textbooks and other educational resources that meet the needs of our unique course offerings. In particular, there are very few books written on diesel-electric locomotives that are applicable for undergraduate engineering students. This book addresses that gap in the literature, and Walter's timing could not have been better.

Across the railroad industry, vast amounts of knowledge are walking out the door in a massive wave of retirements. As railroad managers with 30–40 years of expertise in locomotive operation and maintenance are leaving the industry, railroads are searching for young, bright engineering students to eventually take their place managing the nearly 30,000 locomotives owned by railroads across North America.

At the same time, locomotive technology is rapidly changing through the implementation of positive train control and the drive for increased safety and efficiency made possible by on-board diagnostics, advanced condition-based maintenance strategies, and more sophisticated locomotive energy management systems. The importance of fuel economy is undeniable when fuel costs have, on average, represented about 20% of the annual operating expenses of Class I railroads (totaling over $95 billion spent by all Class I railroads from 2006-2015).

Railroads and policy makers are growing increasingly interested in reducing pollution and fossil fuel consumption.

Walter has done an excellent job combining his already extensive knowledge on the subject with additional insights and information from many locomotive experts in the railroad industry. I look forward to using Walter's *Diesel-Electric Locomotives* as a text for our Railroad Mechanical Practicum course. The book is a good fit because, like Walter, many engineering students today are not only interested in how diesel-electric locomotives work but also are extremely passionate about energy and energy conservation. As the "green" revolution advances across the globe, the railroad industry will benefit greatly from this new generation of railroad engineers who can uniquely address the challenges of safety, efficiency, and sustainability.

Walter's book should well serve educators and students in railroad engineering and other academic disciplines, railroad industry practitioners, academic researchers, and government officials. Additionally, railroad enthusiasts across the globe will find welcome information, photographs, and illustrations in this book as they seek to understand and appreciate the inner workings of these widely admired "diesel workhorses."

Bryan W. Schlake
Altoona, PA
November 2018

Illinois Railroad Museum, September 2016. Photo by author.

Preface – Why Energy?

Welcome to an exploration of diesel-electriclocomotives unlike many you may have seen. While this book is an unabashed appreciation of diesel-electric locomotives, it also examines these technological marvels, component by component, from an energy point of view. It explains in detail how these giant machines work; how they consume and conserve energy; and how they could become more energy efficient and environmentally sustainable.

I chose an "energy and environment" perspective for two reasons. First, I thought it would provide a most revealing tour of the inner workings of these locomotives. And, second, I don't think the "Energy Crisis" is over. In fact, I think just the opposite—it's here and it's accelerating.

Consider the environmental impacts of our continued use of fossil fuels—coal, oil, and natural gas:

- Air pollution from burning fossil fuels
  - Particulates—causing asthma and other respiratory illness
  - Sulfur and nitrogen oxides—causing acid rain, smog, respiratory illness
- Water pollution from oil spills, coal mining, and hydrofracking
- Loss of land and wilderness due to mines, drilling, and pipelines
- Climate change caused by carbon

"I hope my readers find that viewing diesel-electric locomotives through an "energy and environment" lens is technically interesting and personally satisfying. But more than that, I hope we have a shared concern for the future, and a desire to explore ways that locomotives and the railroad industry that operates them can become more environmentally sustainable."

dioxide emissions from burning fossil fuels, which results in:

- Higher global average temperature and generally higher local temperatures
- Increased frequency of extreme heat waves
- Droughts that reduce water supplies and soil moisture—undermining agriculture and contributing to forest fires
- More extreme weather and rainfall events and flooding
- More powerful hurricanes fueled by warmer ocean temperatures, e.g., Hurricanes Harvey (which dropped 54 inches of rain on Houston, Texas) and Maria, which devastated Puerto Rico and other Caribbean islands in September 2017.
- Acidification of oceans and loss of coral reefs and fisheries
- Melting of glaciers, Greenland's ice sheet, Antarctica's ice shelves, and Arctic sea ice
- Rising sea levels, impacting coastlines, coastal cities, and coastal fresh water aquifers and water supplies
- Potential disruption of ocean currents and global thermal transport mechanisms
- Warming of permafrost, releasing trapped CO_2 and methane (another greenhouse gas)
- Potential for additional episodes of colder winter weather in midlatitudes due to disruption of the polar vortex air mass caused by accelerated Arctic warming
- Poleward movement of disease vectors
- Significant acceleration of species extinction

These impacts can be addressed by using fewer fossil fuels through a combination of energy conservation and switching to cleaner energy sources.

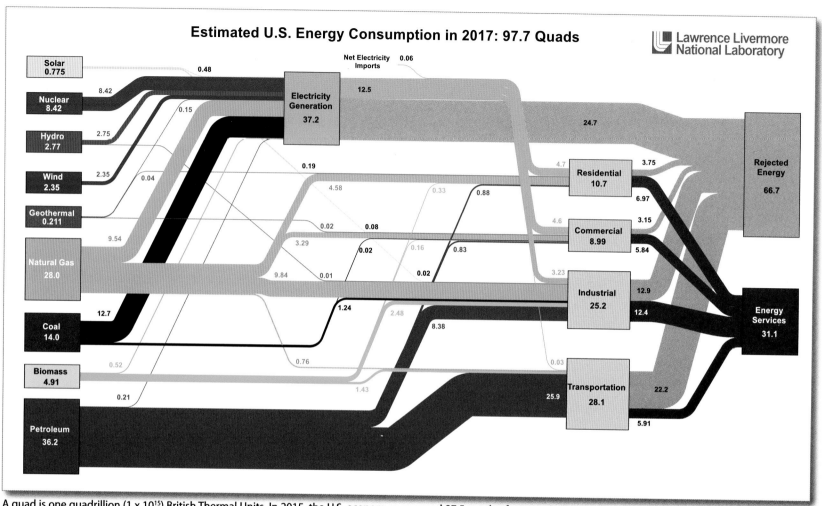

Estimated U.S. Energy Consumption in 2017: 97.7 Quads

Lawrence Livermore National Laboratory

A quad is one quadrillion (1 x 10^{15}) British Thermal Units. In 2015, the U.S. economy consumed 97.5 quads of energy. Image Credit: Lawrence Livermore National Laboratory, U.S. Department of Energy.

The "spaghetti" chart above shows how energy flows through the U.S. economy. The left side of the chart shows the quantities of different energy resources entering the U.S. economy. Note that currently just a small percentage of our energy supply comes from renewable sources. The right side shows both how much energy reaches the point of use *(bottom right)* and how much becomes energy waste or inefficiency *(top right)*.

The spaghetti chart reveals that almost 60% of the energy the United States consumed in 2015 was rejected or wasted! But the percentage of energy waste was actually greater than 60% because that figure does not include the energy wasted by the inefficient devices we use at point of use, or the energy waste associated with unnecessary activities. Thus, the vast majority of the energy we consume is wasted, compounding the environmental and social impacts associated with our over-reliance on fossil fuels. Fortunately, the strategies and technologies for reducing and eliminating energy waste are well known. And, as has been said many times before, "The cleanest and cheapest BTU or kilowatt-hour (kWh) is the one you don't use."

The railroad industry, like any industry, is a microcosm of this spaghetti chart. Each year, Class I railroads[1] alone consume approximately 3.7 billion gallons of diesel fuel annually, and—despite substantial efforts to conserve energy—much of it is wasted. In 2015, their fuel bill was $6.67 billion—though it averaged over $10 billion per year between 2011 and 2015.[2]

The good news is that the railroad industry has a strong commitment to energy conservation and efficiency as well as a history of experimenting with alternative fuels. While saving money has been the primary driver, the industry also expresses a strong environmental commitment through its press releases and annual or biennial sustainability reports, which often reference corporate progress toward greenhouse gas (GHG) emissions reduction goals. The GHGs accumulating in greater concentrations in the atmosphere due to human activity are carbon dioxide, methane, nitrous oxide, ozone, and chlorofluorocarbons.

I hope my readers find that viewing diesel-electric locomotives through an "energy and environment" lens is technically interesting and personally satisfying. But more than that, I hope we have a shared concern for the future, and a desire to explore ways that locomotives and the railroad industry that operates them can become more environmentally sustainable.

* * * * *

While the table of contents lays it all out, I thought it might be helpful to provide a few notes about how this book is organized. So here goes...

Chapter 1, "Introducing Diesel-Electric Locomotives," provides an overview of diesel-electric locomotive technology—its history, significance, and advantages. A brief discussion of Class I railroads' diesel-electric locomotive fleets is also provided. Beyond the scope of this book is a discussion of older locomotive models.

Chapter 2, "How Diesel-Electric Locomotive Energy Systems Work," explains in detail how modern locomotives work —technology-by-technology. This chapter is arranged to *follow the energy flow through the locomotive*—from fuel tank to wheels. Each subsection concludes with a brief summary of energy efficiency opportunities. Discussions of braking systems, with special attention

to dynamic braking, and other energy-related systems are included.

Chapter 3, "Diesel-Electric Locomotive Fuel Economy and Energy Efficiency," delves into the potentially confusing subject of how much energy diesel-electric locomotives use compared to the work they do. Here, the various definitions and conceptualizations of railroad and locomotive fuel economy and energy efficiency are listed and explained in detail. Chapter 4, "Energy Efficient Diesel-Electric Locomotive Operation," then examines strategies for operating these locomotives with maximum fuel economy and efficiency.

Chapter 5, "EPA Emissions Standards, Tier 4 Locomotives, and Energy Impact," shifts the discussion to emissions reduction—the regulations and the technologies for achieving compliance, and their energy impact. This is followed by Chapter 6, "Future Directions for Diesel-Electric Locomotives," which comprehensively scrutinizes technologies and alternate fuels for minimizing the environmental impact of locomotive use. Here special emphasis is given to reducing greenhouse gas emissions to minimize climate change impacts.

Chapter 7, "Railroad Environmental Sustainability," brings the book to a conclusion by briefly surveying (and praising) railroad sustainability efforts and identifying remaining challenges.

While there is discussion of DC locomotives (those with direct current traction motors), the discussion here is primarily "AC-centric"—focused on more modern locomotives with AC motors.

For those who want to know "where that came from," or want to research topics more thoroughly, the book contains a long reading list and hundreds of endnotes for reference. The endnotes are there so the interested reader can make use of them!

I welcome your thoughts and feedback and can be reached via www.diesel-electric-locomotives.com.

I hope you enjoy the book!

Walter Simpson
Amherst, New York
November 2018

BNSF Cicero Yard, Chicago, IL, April 2013. Photo by author.

Acknowledgements

This book has been a pleasure and adventure to research and write —in large part because of all the knowledgeable and pleasant people I have met, talked to, and corresponded with along the way. While I bear full responsibility for the final text, errors and all, I am grateful for the abundance of help many people have so generously offered. The phrase "it takes a village" applies in full. The end result for the reader is a much better, more comprehensive, and, I trust, more accurate book.

I would like to begin by expressing special thanks to Bryan Schlake, Instructor of Rail Transportation Engineering at Penn State-Altoona, who appreciated the potential of this book project from the beginning. He read and commented on the manuscript numerous times, introduced me to industry experts, and was always there to help and guide me as the book progressed. I could not have written this book without his knowledgeable suggestions, good ideas, generous assistance, and encouragement.

I would also like to offer special thanks to Don Graab, Allen Rider, and Dave Cook, who were especially generous in sharing their vast locomotive expertise, reviewing the book manuscript, and offering needed corrections and helpful suggestions. Additionally, and at the risk of leaving someone out, I wish to acknowledge and thank the following individuals for their interest and help in providing technical, literary, research, or production assistance; some also read and commented on early and midstream drafts or portions of the manuscript: Don Anair, Jennifer Bailey, Bruce Becker, Munidhar S. Biruduganti, Joan Bozer, Ian Bradbury, Andrew Burnham, Adam Burns, Bryan Chaisson Sr., Stephen Christy, Stephen Ciatti, Jeff Cutright, Richard (Colby) Davis, Steve Dillon, Fred Dryer, Mark Duve, Amgad A. Elgowainy, Jon Fedora, Claudio Filippone, Marcus Gillebaard, Cody Graham, Sean Graham-White, Ashok Gupta, Christopher Holt, Hugh and Brenda Hopkins, Mike Jaczola, Timothy Juliani, Ron Kamen, Johannes Kech, Pat Keister, Bob Kennedy, Wayne Kennedy, Bill Keppen, Tammy Krause, Elena Krieger, Martha Lenz, William Lisk, Douglas E. Longman, Greg Lund, David Lyon, Tom Mack, Jeremy Martin, Ernie McClellan, Greg McDonnell, Bruce McFarling, Jason Miller, Mike Miller, Dave Reichmuth, Paul Rhine, Erin Ryan, Joshua Sams, Chris Saricks, Paul Schiff, Don Scott, Walter Talbot, Gerhard Thelen, J. Craig Thorpe, Graciela Trillanes, James Ulrich, William Vantuono, and Brian Yanity.

I offer special appreciation to Paul Rhine. I loved his book, *Fuel Saving Techniques for Railroads: The Railroaders Guide to Fuel Conservation*, in a way only an energy guy could. It was an early source of motivation. The *Railroad and Locomotive Technology Roadmap*, prepared by the U.S. DOE's Argonne National Laboratory, was also an early discovery that motivated me, and I thank the government and industry contributing authors of this wonderful document, written in 2002 but somehow not dated. While perhaps a little off-topic, I also wish to acknowledge and thank Karen Parker for the very helpful *How a Steam Locomotive Works* which unraveled some technical puzzles for me about steam locomotive technology while planting seeds for my book. And local thanks to Jim Ball and the Niagara Frontier Chapter of the National Railroad Historical Society for giving me the opportunity to present my "Energy and Trains" talk at a monthly meeting a few years ago—an event which helped galvanize the idea of writing this book.

Because locomotive manufacturers and railroads generally don't talk about diesel-electric locomotive energy efficiency, investigating that topic was at times like peeling back an onion—or what one consultant called penetrating the *cone of silence*. I wish to thank the railroad industry experts who helped me in this exercise including those who wished to be anonymous.

I also wish to thank the photographers, illustrators, railroads, manufacturers, and other creators or owners of the images that grace the pages of this book. They are credited under each image and were kind and generous in granting me permission to include their works here. Special thanks to designer/illustrator Robert Hochberg, *Trains* magazine/Kalmbach Publishing (Diane Laska-Swanke), EMD/Progress Rail/Caterpillar (Barbara Jansen), and Norfolk Southern (Jennifer McDaid, Marc L. Orton and Sharde Shorts). These images have improved the book immeasurably, as did those taken at Norfolk Southern's Juniata Locomotive Shop in Altoona, PA.

Special thanks, too, to my daughter Skye Simpson, whose professional writing talents and good nature allowed her to serve skillfully as my in-house copy editor, and to Pam Rose for her tireless efforts locating one technical study or reference book after another through interlibrary loan and the University at Buffalo's library system.

I also offer special thanks to Brian Brundige at Simmons-Boardman Books for his support and interest in my project. Brian has been a pleasure to work with. He always responded positively and supportively, while turning my manuscript into what I hope readers believe is an excellent publication. My hat is also off to his talented staff—Linda Baker, Cindy Gottlieb, and Tom Leary—for their hard work and contributions to the final product. Previously mentioned Robert Hochberg deserves additional thanks for working with the Simmons-Boardman team to design this book.

Trains and railroads have touched my life in many ways. I am fortunate to have a family that loves trains, from my aunt Rita Collmar, who traveled every weekend between Philadelphia and Maplewood, New Jersey, to visit me as a baby, to my brother Ron Simpson who still gladly takes me to Amtrak's Northeast Corridor every time I visit him in Wilmington, Delaware, and to my sister Judie Simpson who introduced me to the Chicago "L" and Metra railroads and encouraged me many times to "Write That Book!" Interest in rail transportation was instilled in me as a young boy when my grandfather took me riding on Philadelphia's trolleys and elevated lines (he was blind and I served as his eyes as we navigated the city) and when my parents got me my first train sets.

Last, and certainly not least, I especially want to thank my wife Nan. I am very fortunate we share a love of trains, but, much more than that, I am so grateful to her for supporting me body and soul through my three-year obsession researching and writing this book. She put up with a lot, but her love and support never wavered. I dedicate this book to her with much love.

Walter Simpson
Amherst, NY

Nan and Walter Simpson
Photo credit: Hugh Hopkins.

Thurmond WV, June 2014. Photo by author.

CHAPTER ONE

Introducing the
Diesel-Electric Locomotive

Diesel-electric locomotives come in different shapes and sizes. Smaller, less powerful "switchers" are primarily used for sorting cars in yards and for short-hauling assignments, while larger, more powerful "road locomotives" (also called *line-haul locomotives*) move freight and passengers over distances between terminals and stations. This book is primarily about road locomotives.

Contemporary diesel-electric road locomotives are very large machines. While riding on standard gauge track (with just 4 feet 8.5 inches between the inner faces of the rails), these locomotives—typified by the GE ES44AC and EMD SD70ACe—are more than 10 feet wide and nearly 16 feet high.[3] They stretch over 70 feet in length and weigh as much as 432,000 pounds. Locomotives are also smart machines. The manufacturers and railroads are increasingly adding new sensors and other technology to allow locomotives to perform on-board monitoring, self-diagnostics, and electronic reporting—which improve both reliability and efficiency.

1.1 Diesel Engine + Electric Transmission

Diesel-electric locomotives have been described alternately as diesel locomotives with "electrical transmissions" or as electric locomotives "carrying their own power plants." These locomotives are undeniably diesel-electric generating plants on wheels because they have large diesel engines (called *prime mover*) that power electric generators. The electricity that is produced runs electric traction motors that turn the locomotive's drive wheels. A modern 4,400 horsepower locomotive can produce enough electricity to power a small town.

Norfolk Southern ES40DC 4000 horsepower locomotive approaching Cresson, PA, 2014, dwarfing our camper. Photo by author.

Locomotive Energy Transformations

The numerous energy transformations within a diesel-electric locomotive begin when a pump draws **chemical energy** in the form of diesel fuel from the locomotive's fuel tank and sends it to the diesel engine for combustion. There, fuel explosions in each cylinder create **thermal energy** (heat). That heat energy is transformed into **mechanical energy** when the explosions in each cylinder push pistons and turn the crankshaft. When this shaft turns the rotor in the locomotive's generator, **electrical energy** is produced, which powers the electric traction motors that produce the **mechanical energy** (also called *kinetic energy*) propelling the locomotive and its train.

We can visualize diesel-electric locomotive energy transformations as follows:

Chemical Energy → Thermal Energy →
Mechanical Energy →
Electrical Energy → Mechanical Energy

Each energy transformation involves energy losses. Minimizing these losses would improve the locomotive's energy efficiency.

Locomotive Schematic
This conceptual diagram shows the key components and general layout of a diesel-electric locomotive

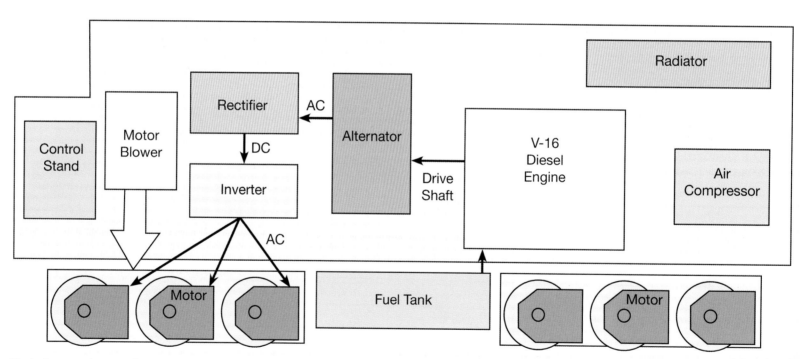

Simplified schematic diagram of a modern diesel-electric locomotive with inverters and AC motors. Note that while new locomotives are equipped with AC technology, most operational locomotives still use DC motors. Locomotive diesel engines may have 12, 16, or 20 cylinders. Image Credit: Nolelander (Wikimedia Commons).

1.2 Advantages of Diesel-Electric Locomotives

When diesel-electric locomotives were coming into their own in the 1930s, 40s, and 50s, eventually replacing steam locomotives, the process was known as *dieselization*. Dieselization was unstoppable because diesel-electric locomotives were superior performers and less expensive to operate than steam locomotives.[4]

The following is a list of diesel-electric locomotive attributes that also describe the advantages these locomotives had over steam locomotives. Many of these attributes, as applied to contemporary diesel-electric locomotives, are discussed in greater detail later in this book.

Acceleration and Heavy Hauling. Diesel-electric locomotives produced maximum tractive effort at low speeds due to the inherent characteristics of series-wound direct current (DC) electric motors, and thus excelled at starting heavy trains. While locomotive experts in the 1940s pointed out that modern steam locomotives of the time could still accelerate passenger trains quicker than diesel-electrics at speeds above 40 mph,[5] and that some steam locomotives could develop very high levels of horsepower,[6] these attributes were less important than the diesel-electric locomotive's dramatic increase in pulling power at low speeds.

Increased Adhesion. Diesel locomotives provided increased adhesion to the rail. Because they can apply power smoothly, they can successfully apply very high levels of tractive effort to the rails without slipping, as is required to get heavy trains moving and to pull them up gradients or hills.

Superior Braking. In addition to the locomotive "independent brakes," with their friction braking against the wheel treads (the portion of the wheel that makes contact with the rail head), diesel-electric locomotives can use dynamic braking that is produced when the locomotive's electric traction motors are

Dieselization is depicted as a major step forward for the railroads in print ads from the 1940s and 50s.

Image Credit: Pennsylvania Railroad (from the author's collection).

Image Credit: © Norfolk Southern Corp.

operated as generators. In this mode, the train's mechanical energy is converted to electrical energy, which is then dissipated as heat in dynamic brake grids located in a compartment behind the locomotive cab or at the rear of the locomotive. One of the advantages of dynamic braking is that it does not impart heat into railcar wheels, which can make them more susceptible to failure. It also extends the life of the brake shoes that press against the wheel treads with conventional air brake applications. (For more on diesel-electric locomotive braking systems, see Section 2.11.)

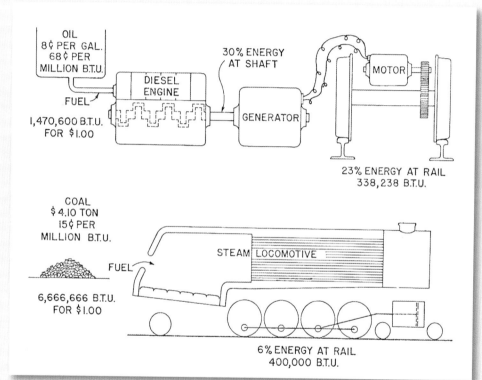

Steam beats Diesel? This illustration suggests so. The Norfolk and Western (N&W) Railway was one of the last major railroads to undergo dieselization. Here we see a late 1940's N&W comparison of fuel-to-rails energy dollar efficiency of diesel-electric and steam locomotives. At diesel fuel and coal prices of the time (8 cents per gallon and $4.10 per ton), $1.00 worth of diesel fuel would deliver 338,238 BTUs of energy to the rails while $1.00 worth of coal would deliver 400,000 BTUs to the rails. Thus, steam won this comparison. Eventually, all the other benefits of diesel-electric locomotive technology caught up with the N&W and new leadership quickly dieselized this coal-rich railroad. Note: The assumed fuel-to-rails energy efficiency of diesel-electric locomotives was almost four times that of steam locomotives—23% vs. 6%. Image Credit: Norfolk & Western Historical Society.

Energy Efficiency. Diesel-electric locomotives are highly energy efficient. While the diesel-electric locomotives that replaced steam locomotives had peak energy efficiencies near 25%, modern diesels can now achieve efficiencies of 35% or more. That means that 35% or more of the energy contained in the locomotive's diesel fuel can be delivered to the rails.[7] Steam locomotives were, at best, 7% to 8% efficient. By comparison, conventional (non-hybrid) automobiles and light trucks are 14% to 30% efficient and hybrid cars are 25% to 40% efficient.[8]

Relatively Non-Polluting. Pollution is a relative term. Diesel locomotives burn diesel fuel, and the end result is pollution in the form of air emissions. Though cleaner than coal-burning steam locomotives, early diesels produced significant levels of pollutants including particulate matter. However, ultra-low sulfur diesel fuel and advances in diesel engine technology such as electronic fuel injection, in response to increasingly stringent U.S. Environmental Protection Agency (EPA) emissions standards, have produced much cleaner diesel engines.

Easier to Fuel. Diesel-electric locomotives don't need giant coaling stations or numerous coal shipments to supply these stations. While they require their own fuel infrastructure, it is far less burdensome and costly. Contributing to the relative ease of fueling diesel-electric locomotives is their ability to go further between refueling stops due to the higher energy density of diesel fuel compared to coal, plus the diesel locomotive's higher energy efficiency. (*Note:* Not all steam locomotives burned coal. For example, many Southern Pacific and Union Pacific steam locomotives burned fuel oil.)

No Need for Large Amounts of Water. Steam locomotives consumed vast quantities of water as well as coal or oil. For example, a large steam locomotive could consume 30,000 gallons of water in its tender while pulling a train just 250 to 300 miles.[9] Obtaining that water was difficult in some areas. Moreover, the water had to be chemically treated to prevent mineral buildup in steam boilers. Water supplies also had to be heated in the winter. In contrast, diesel-electric locomotives require 300-400 gallons of water for engine cooling purposes, and it is continuously recycled with few, if any losses. Note that both steam locomotives and early diesel-electric locomotives consumed water to provide passenger cars with steam for space-heating purposes. Today, passenger cars are heated by 480-volt electricity provided by the locomotive, and this is referred to as "head-end power" (HEP) or "hotel power." HEP systems also power lighting and air conditioning.

An early GM Electro-Motive Diesel full-page print advertisement touting the benefits of EMD locomotives and dieselization generally. Courtesy of former General Motors Electro-Motive Division, now part of Progress Rail, A Caterpillar Company.

Standardized, Mass-Production. Unlike steam locomotives, diesel-electric locomotives became standardized (in terms of fewer models) and were mass-produced efficiently with standard parts on an assembly line. Standardization also made maintenance easier and reduced both production and maintenance costs—though early diesel-electric locomotives were more expensive than the steam locomotives they replaced.

GM EMD shop output described as the "largest production of new railroad horsepower in the world," April 14, 1951, *Saturday Evening Post*. Courtesy of former General Motors Electro-Motive Division, now part of Progress Rail, A Caterpillar Company.

Low Maintenance. Steam locomotives were historically serviced at 100- to 120-mile intervals (at crew change points). It was only during the later years of steam, when steam locomotive technology evolved to use multiple-point mechanical lubrication systems and roller (cylindrical) bearings, that service intervals on railroads such as the Norfolk & Western were extended. In contrast, while diesel locomotives required routine maintenance and periodic overhauls, they were inherently less maintenance-intensive than steam locomotives. Moreover, over the years, locomotive designers made strides in reducing the maintenance requirements of diesel-electric locomotives, although the latest generation of emissions-compliant engines appear to be reversing that trend.

High Availability. While steam locomotives were generally available for service 60% of the time[10] early diesel-electrics could achieve better than 80%. One authoritative source, ascribing diesel-electric availability as high as 95% in 1946, argued that the superior availability of diesel-electrics was the primary reason railroads switched to them.[11] High availability ultimately allowed railroads to provide equal or better service with fewer locomotives. Modern diesel-electric locomotives are available for service 90% to 95% of the time.

Longevity. Diesel-electric locomotives last a very long time. For example, it is not uncommon for railroad companies, especially smaller railroads, to operate 40-year-old diesel-electric locomotives in regular service. Moreover, during their operational lives, some diesel locomotives may travel as much as 1 million miles or more.[12] This, of course, is a plus/minus proposition, the minus being that very old locomotives are much less energy efficient. Some steam locomotives had long careers, too, but the number of miles they traveled over the course of their

F Type locomotive production at General Motors' EMD plant in La Grange, Illinois, circa 1945. Courtesy of former General Motors Electro-Motive Division, now part of Progress Rail, A Caterpillar Company.

lifespan was generally much less. One exception was the Norfolk & Western Railway's 4-8-4 J Class passenger steam locomotive. The J's averaged 15,000 miles per month, with several locomotives recording in excess of 3 million miles before retirement in 1959.[13]

Modern Symbolism. While steam locomotives were viewed as dirty but loyal servants of the past, diesel locomotives —

Norfolk & Western J Class steam locomotive, several of which exceeded 3 million miles of service. Petersburg, VA, 1952. Courtesy of Norfolk & Western Historical Society.

especially streamlined models—were symbols of technological progress and modernity. The addition of colored livery to these new locomotives added to their appeal. While this symbolism exists less today, new locomotives—electric and diesel-electric—remain visually impressive.

Less Rail Damage. Due to the crudely balanced nature of steam locomotive drive wheels and these locomotives' uneven application of force to the rail, steam locomotives were notorious for "rail-pounding" and resultant damage to rails and track structure. In contrast, diesels applied force smoothly through the operation of electric traction motors—minimizing the need for "section gang" track workers.

Easier and Safer to Operate. Diesel locomotive controls are simpler and easier to operate than were those in a steam locomotive, and operational mistakes do not result

Chicago–New York (Jersey City) Lake Cities preparing to depart Dearborn Station, Chicago, early 1950s. Photo Credit: Wallace W. Abbey.

in boiler explosions. Diesel cabs are increasingly more comfortable for crews, and the location of the cab in the front of the locomotive increases visibility for safe operation.

Multiple-Unit Operation. Unlike steam locomotives, multiple diesel-electric locomotives can be operated by one engineer in the lead locomotive. The secondary locomotives

"Huskies and Their Equipment," Pennsylvania Railroad Blacksmith Shop Workers, Altoona, PA, 1926. Photo Credit: Walter P. Keely, Jr.

than steam. So, this advantage or "plus" for diesels was also a big negative for railroad workers, and their families and communities. While improving the national economy and the financial viability of railroads, dieselization was destructive to the local economies of many towns and cities, some of which were founded by railroad companies, e.g., Altoona, PA, by the Pennsylvania Railroad in 1849. The human toll was immense.

Cost Saving, Profit Boosting. Diesel-electric locomotives cost more than steam locomotives of equal horsepower, but many of the advantages of diesel-electric locomotives translated into lower overall costs with greater productivity. The cost savings came in many forms:

- Reduced fuel and fuel system costs
- Reduced water and water system costs
- Reduced locomotive maintenance and downtime
- Number of locomotives
- Reduced number of crews
- Reduced need for helper locomotive service in mountainous districts
- Reduced damage to track
- Reduced brake shoe replacement cost (due to dynamic braking)

The bottom line: Diesel-electric locomotives reduced costs, and, in so doing, produced higher profits for railroad companies and their owners.

with interconnected controls can be located immediately behind the lead locomotive, at the end of the train, or in the middle. Diesel-electric locomotives with multiple-unit (MU) operation capabilities allowed longer trains without running "second sections" or "double-heading." A *second section* is a closely following second train whose operation is necessary because the full load could not be handled by the first section. *Double-heading*

means running two locomotives at the head of the train—each with their own crew.

Far Fewer Workers. If you worked for a railroad or hoped to work for one during the transition to diesel-electric locomotives, that transition must have seemed like the end of the world because the construction, maintenance, and operation of these diesel locomotives required far less manpower

1.3 Critical Energy Systems for Diesel-Electric Locomotives

The diesel-electric locomotive was made possible over a century ago by the invention and perfection of two key technologies—namely, the diesel engine itself and its "electrical transmission." The latter refers to the drivetrain, which converts mechanical energy from the locomotive's diesel engine into electrical energy that's put to work powering electric traction motors that deliver mechanical energy to the rails. These motors are called *traction motors* because they are used to provide tractive force for vehicle propulsion. In a modern diesel-electric locomotive, the electrical transmission consists of the traction alternator, rectifiers, inverters, traction motors, gearing, and electrical switch gear.

Rudolph Diesel and his first engine. Photos courtesy of Dieselnet (public domain).

Diesel Engine. In 1893, Rudolf Diesel (1858–1913) published *Theory and Construction of a Rational Heat-Engine to Replace the Steam Engine and Combustion Engines Known Today* and received his first patents for the "compression-ignition engine" or what we now call the *diesel engine*.

Unlike a steam engine, which is an *external* combustion engine, a diesel engine is an *internal* combustion engine. As such, it achieves greater levels of energy efficiency by directly using the force unleashed by fuel combustion to create mechanical energy. Diesel engines are noted for compression ratios above 14:1,[14] where *compression ratio* is the ratio between the largest volume in an engine cylinder (when the piston is at its lowest point) and the smallest volume (when the piston has reached the top of its stroke). Heat generated by high compression in diesel engines causes the fuel to self-ignite. Spark plugs are not needed.

Diesel engines are especially energy efficient because they operate at high pressure with a "leaner" oxygen-rich fuel mix.

Diesel built his first operational engine in 1897, achieving 26% efficiency. While still much lower than what Diesel believed was theoretically possible, this efficiency compared very favorably to the 7% to 8% efficiency a steam locomotive could achieve under optimal circumstances. Diesel's compression-ignition engines could also operate on a variety of fuels, including powdered coal and vegetable oil.

Early diesels were large, heavy, four-cycle engines, suited for stationary or marine use. The development of two-cycle diesel engines, which were lighter than four cycle engines of similar output, was of critical importance to adaption of diesels to railroad locomotives.

This CSX GE AC4400CW diesel-electric locomotive is powered by a GE 7FDL-16 engine. Each cylinder in this 4-cycle V-16 engine displaces almost 700 cubic inches. New River Gorge, WV, 2013. Photo by author.

Four-Cycle vs. Two-Cycle. A four-cycle diesel engine (also referred to as a *four-stroke engine*) has one power stroke for every four strokes, which consists of two up-and-down piston movements or two revolutions of the crankshaft. In this engine, there is a power stroke, followed by an exhaust stroke, followed by an intake stroke, followed by a com-pression stroke, at the end of which the diesel fuel is forcefully injected into the cylinder. It then self-ignites, initiating the next power stroke.

In contrast, a two-cycle diesel engine has one power stroke for every two strokes, which consists of one up-and-down piston movement or one revolution of the crank-shaft. While the compression stroke, injection event, and power stroke in two- and four-cycle diesel engines are similar, in a two-cycle engine the exhaust and intake functions are combined together and take place when the piston is near bottom dead center of its power stroke and in the first inch of its compression stroke. Here is how that works: When the piston is near the bottom of its power stroke, exhaust valves located in the top of the cylinder open, allowing expanded combustion products to begin exiting. Then, in a process called *scavenging*, inlet ports in the cylinder wall near the bottom of the cylinder open (by being uncovered by the piston as it moves down in the cylinder), allowing pressurized combustion air to enter the cylinder and force out remaining exhaust gases through the exhaust valves. As the piston starts to move up in the cylinder, the exhaust valves close, then the inlet ports close by being covered by the piston as it moves up. Simultaneously, the compression stroke begins. Near the top of the compression stroke, diesel fuel is injected into the cylinder and combusted by the heat of compression.

While two-cycle engines have disadvantages (e.g., emissions due to oil consumption), they respond quickly and have higher "power density" (power-to-engine size and weight) ratios than four-cycle engines because they have twice as many power strokes for any given rate of revolution (rpm). Friction per power stroke is cut in half. And because (a) horsepower is a function of Pressure x Cylinder Bore x Piston Stroke, and (b) two-cycle engines have an effective stroke twice the length of four-cycle engines, two-cycle engines can theoretically operate at lower pressures or lower speeds and produce the same horsepower as four-stroke engines. This potentially improves reliability. However, as a general rule, four-cycle diesel engines are more energy efficient because they more effectively intake combustion air and exhaust combustion products.

Electric Transmission. We think of a hybrid automobile—with sophisticated technology combining the operation of a gasoline engine, electric generator, electric motors, and batteries—as a new invention, but early hybrid cars with gas-electric transmissions were invented and in use 100 years ago. They were called *Dual Power* cars and were specifically designed to achieve optimal fuel efficiency by operating an internal combustion engine at full load and then using the portion of its output not immediately needed to propel the car for battery charging.

To meet driving conditions efficiently, dual power cars were able to operate in a number of modes—all gas, all electric, or a combination of gas, electric motor, and battery outputs. They even had regenerative braking—their electric traction

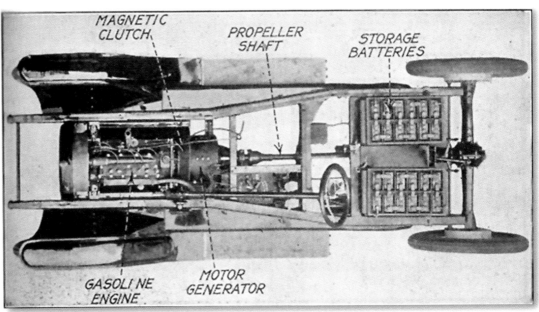

Diagram of hybrid drivetrain of 1917 Woods Motor Vehicle Company "dual power" car. Image Credit: Woods Motor Vehicle Company (Public Domain).

motors would provide a braking force by operating as generators when the dual power car was coasting downhill or coming to a stop. The electricity produced by regenerative braking recharged the car's battery. The author's Prius C hybrid car just having achieved 75 mpg (miles per gallon) on a round trip from Buffalo to Niagara Falls demonstrates the cleverness of the technology. It was all the more so 100 years ago!

The marriage of an electric transmission to an internal combustion engine in a locomotive has a long and interesting history.[15] While it is not possible to do justice to that story here, a few items can

be mentioned. In 1888 and 1889, Charles Bradley of Yonkers, New York, patented the first gas-electric transmission system in the United States. Similar patents were granted to William H. Patton of Chicago in 1889 and 1892. Other early patents for an electrical transmission as applied to a locomotive were granted to Martin E. Cassady of Kentucky in 1898. But it was not until 1914, when Herman Lemp (1862-1954) of Ridgewood, New Jersey, invented and patented electrical controls capable of effectively coordinating engine and generator outputs, that gas-electric locomotives and railcars (the latter known as *doodlebugs*[16]) became practical. Lemp worked for

"Dawn to Dusk Club" passengers who rode the *Pioneer Zephyr* during its record-setting run from Denver to Chicago on May 26, 1934. Photo Credit: Chicago Tribune (Public Domain).

they gained national attention when Chicago, Burlington & Quincy's lightweight diesel-electric *Zephyr* streamliner broke railroad speed records by traveling nonstop from Denver to Chicago in 13 hours and 5 minutes.

In 1939, General Motors' Electro-Motive Division created the F-Type freight locomotive. FT locomotive #103 was credited by David P. Morgan, long-time editor of *Trains* magazine, as "The Diesel that Did It," i.e., the diesel-electric locomotive that sealed the fate of steam locomotives, hastening the inevitability of dieselization.[18]

In a brilliant marketing move, beginning in 1939 this brand-new four-unit 5,400 hp diesel locomotive traveled 83,764 miles, crisscrossing the United States to demonstrate its capabilities to various railroads. According to Morgan, it had "no equal in railroading."[19] This demonstration was followed by other General Motors' public relations campaigns, including one in 1947 when GM's "*Train of Tomorrow*" began touring the country. This train featured a 2,000 hp E7 GM diesel locomotive pulling four Astra Dome cars. It logged an impressive 65,000 miles showing off its futuristic features.

The history of diesel locomotives is rich and beyond the scope of this book. Many companies besides GE and EMD made major contributions to the development

General Electric, which—along with Westinghouse—provided electrical equipment to builders of early electric and gas-electric locomotives.[17]

Combining the diesel engine with an electric transmission in a railroad locomotive was not easy because the four-cycle diesel engines of the era with suitable output were very large and heavy. Nonetheless,

by 1924, GE, Ingersoll Rand, and the American Locomotive Company (ALCO) created functional diesel-electric locomotives with four-cycle diesel engines. These proved the concept, and at least one operated as a yard switcher for many decades.

During the ensuing decades diesel-electric locomotives proved themselves slowly but surely. In 1934, for example,

It's years ahead of schedule

General Motors' *Exciting New* *Train of Tomorrow*

Photo on left: Publicity photo for the GM *Train of Tomorrow* Milwaukee, April 26, 1948. *Image on right:* vintage General Motors print ad for "a new and wonderful train". Photos Courtesy of former General Motors Electro-Motive Division, now part of Progress Rail, A Caterpillar Company.

of diesel engine and diesel-electric locomotive technology. These included ALCO, Baldwin, and Fairbanks-Morse.

It's important to note that the electric transmissions of the Ingersoll Rand/ALCO boxcab locomotive, the *Zephyr*, and generations of U.S. diesel-electric locomotives since then have not had the storage batteries and regenerative braking characteristic of dual-powered cars in 1915. After all these years, we are still waiting for production model hybrid diesel-electric locomotives to arrive.

Electric Transmission Configurations. The configuration of diesel-electric locomotive electric transmissions has evolved over time, as generator and traction motor types changed. Early diesel-electric locomotives had direct current (DC) generators powering DC traction motors, so their electric transmissions were configured along these lines:

Diesel Prime Mover→Generator → ***Direct Current****→DC Electric Traction Motors*

Around 1966, alternating current (AC) alternator-type generators replaced DC generators in diesel-locomotives when solid-state diode technology progressed to provide compact reliable "rectification" for new production locomotives. Rectification converted AC to DC. These locomotives were configured as follows:

Diesel Prime Mover→Alternator→ ***Alternating Current****→Rectification→* ***Direct Current****→ DC Traction Motors*

Diesel-electric locomotive electric transmission systems became more complicated in

Understanding Force, Energy, Power, and Velocity

In various places, this book describes diesel-electric locomotives and their operation in terms of force, energy, and power. Here are definitions of these terms and their interrelationships.

Force

Force is applied with a push or pull. In our discussion, an amount of force is given in units of pounds, typically in the context of the tractive force (called *tractive effort*) a locomotive can deliver to its wheels for propulsion – when starting and in continuous operation. Formally, force is defined as mass multiplied by acceleration:

$$Force = Mass \times Acceleration$$

Energy

In the language of mechanical engineers and physicists, *energy* is equal to *work*. Energy quantities are given here in units of British Thermal Units (BTUs), pounds or gallons of fuel, and horsepower-hours (hp-hrs). Formally, energy is force multiplied by distance:

$$Energy = Force \times Distance$$

Power

Power is the rate of energy production or consumption. Quantities of power are given here in units of horsepower or megawatts. Formally, power is energy divided by time:

$$Power = Energy/Time$$

Power, Force, and Velocity

Railroads can calculate locomotive tractive force or effort by dividing the power a locomotive is delivering to the rails by its speed.

$$Power = Energy/Time = (Force \times Distance)/Time = Force \times (Distance/Time)$$

Thus, *Power = Force × Velocity*

If *Power = Force × Velocity*, then *Force = Power/Velocity*

Thus, force and velocity are inversely related. When one increases, the other decreases, and vice versa. When locomotive power output is held constant, the force a locomotive can apply to the rails keeps decreasing as the locomotive's velocity increases. Eventually, when tractive force has declined to a point where it is equal to the total train resistance, the locomotive will be unable to accelerate unless its power output increases. If it's at maximum power output, that speed will be its top speed.

Assuming power is given in units of horsepower and speed in units of miles per hour (mph), a correction factor of 337.5 can be used to produce force in terms of pounds.

Tractive effort in pounds = 337.5 x horsepower at rails/ velocity in mph

Traditionally, a correction factor of 307.5 has been used, which assumes that 82% of the power generated by the diesel engine makes it to the rails. Newer locomotives are more efficient in delivering engine horsepower to the rails. The correction factor of 337.5 assumes that 90% of engine horsepower is delivered to the rails.

To move the train, the tractive effort applied to the rails by a locomotive must overcome the train resistance (which is comprised of journal bearing resistance, flange resistance, and air resistance), grade resistance, and curve resistance —all measured in pounds of force.

the 1990s when these locomotives began using AC electric traction motors with their multiple benefits (e.g., better adhesion and improved energy efficiency). Controlling AC motors required the use of an inverter, an electrical device that inputs DC and outputs AC. Locomotive inverters are designed for variable frequency and variable voltage output – just what is required to control AC traction motors. Thus, the diesel-electric locomotive's electric transmission evolved into:

Diesel Prime Mover→ Alternator→
Alternating Current→*Rectification→*
Direct Current→*Inverters→***Alternating Current**→*AC Traction Motors*

The line-haul or road diesel-electric locomotive in this configuration, i.e., with inverters and AC traction motors, is the primary subject of this book. The energy systems of

1.4 Performance Criteria for Diesel-Electric Locomotives

that type of locomotive will be discussed in detail in coming chapters. Some discussion of DC locomotives is included as well. What makes a locomotive a good performer and attractive to a railroad company? While this book is primarily concerned with energy issues as they pertain to locomotives, performance has many dimensions and railroads consider many criteria when determining the value of a locomotive. Locomotive manufacturers sell their products based on these features and others:

- **Traction horsepower**—The amount of horsepower a locomotive produces for traction purposes, typically as delivered to the traction alternator; greater horsepower is essential for faster acceleration, higher speeds, and pulling heavier trains.
- **Starting and continuous tractive effort**—The maximum amount of effort or force, measured in pounds, that a locomotive can deliver through its wheels to the rails under ideal circumstances. It is used to start a train and then on a continuous basis to move the train forward.
- **Adhesion**—The ability of a locomotive to apply tractive effort to the rails without wheel slippage. This is a function of a locomotive's "adhesion factor" (tractive effort divided by the locomotive's weight on its driving wheels), traction motor type, motor controls, truck design, and anti-slip technologies.
- **Fuel economy or energy efficiency**—The rate of fuel consumption divided by horsepower output. It is also expressible as a percentage, calculated by dividing work output by energy input.
- **Responsiveness**—This is the rate at which a locomotive responds to changes in throttle settings.
- **Maximum speed**—While higher top speed capability is a consideration for passenger diesel-electric locomotives, freight road locomotives are geared for a maximum 70 to 75 mph and typically operate at slower speeds.
- **Ride quality and handling**—Rough riding locomotives present safety concerns, wheel and track wear problems, and crew comfort issues.
- **Maintenance**—Minor repairs and periodic checkups and overhauls are expected but not excessive "maintenance events."
- **Availability and dispatch reliability**—Railroads want locomotives that show up for work and do not breakdown; they don't want locomotives that are unavailable or fail *on the line of road* (en route potentially blocking other trains).
- **Longevity**—It is the likely locomotive operational lifespan.
- **Emissions**—Locomotive should be compliant with EPA emissions standards.
- **Financial costs/benefits**—These include locomotive purchase price, lifecycle costs, productivity, and earnings potential.

While locomotives under consideration for purchase must meet railroad specifications for all of the above features, the "financial cost/benefit" of a prospective locomotive over its expected lifespan is most important when purchasing decisions are made. Purchasing executives know that buying a locomotive means being wedded to it for decades and they expect it to make money for their company over that period.

1.5 Today's Diesel-Electric Locomotive Fleet

The U.S. diesel-electric locomotive fleet contains over 38,500 locomotives, with approximately 95% in freight service and 75% belonging to the seven Class I freight railroads. The average age of all operational U.S. locomotives is nearly 24 years. Over 50% have diesel engines or prime movers with 4,000 traction horsepower or more. While locomotives with AC traction motors represent nearly 100% of new sales, two-thirds of operational diesels have DC motors.[20]

General Motor's Electro-Motive Division (EMD) dominated new diesel-electric locomotive sales during the 1940s, 50s, 60s, 70s, and 80s until 1987 when General Electric diesel locomotives sales began surpassing EMD.[21] GE has maintained a sales advantage ever since.

GM's EMD became "Electro-Motive Diesel" when purchased from GM in 2005 by the Greenbriar Equity Group and Berkshire Partners, which then sold the company to Progress Rail Services Corporation, a subsidiary of Caterpillar, in 2010.

Locomotive production is highly cyclic. There can be wild swings in the number of orders. Impressive sales figures one year may be followed by a steep reduction in sales for a number of years. For example, in late 2017 neither GE or EMD had any new domestic orders, with Class I railroads having hundreds of locomotives idle and stored—awaiting reactivation or rebuilding. Then, in 2018, some Class I railroads announced capital budgets that earmarked new locomotive investments, while GE and EMD announced contracts for overseas sales.

When reductions in new orders lead to layoffs, the builders lose talented and skilled employees, which makes rebounding difficult. While competition is essential, the U.S. market was not large enough to support three freight locomotive manufacturers (witness ALCO's departure in 1969), and the two remaining builders, GE and EMD, struggle in times of economic downturn.

Describing the future of transportation as less promising than its other businesses,[22] GE sent shockwaves through the railroad industry when it announced in November of 2017 that it planned to sell GE Transportation, its locomotive manufacturing division.

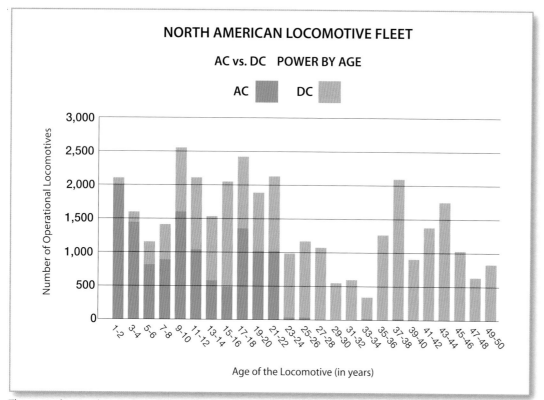

The vertical axis is the number of locomotives and the horizontal axis is the age of those locomotives. New diesel-electric locomotives have AC traction motors. 2015 data. Image Credit: Railinc Corporation.

This followed a July 2017 GE announcement that it planned to transfer locomotive manufacturing from its Erie, PA, plant, where it has produced locomotives since 1910, to its Fort Worth, TX, plant. As of May 2018, GE planned to sell its locomotive division and other segments of GE Transportation to Wabtec Corporation, a locomotive equipment supplier and builder formed through a merger of Westinghouse Air Brake Technologies Corporation and MotivePower Industries.[23]

On the U.S. freight locomotive side, the most technologically sophisticated diesel-electric locomotives today are the U.S. EPA Tier 4-compliant GE ET44 series and EMD SD70ACe-T4 locomotives. The ET44 locomotive was available for sale in 2015, while first deliveries of the SD70ACe-T4 were delayed until 2017. Both of these locomotives have 4-cycle V-12 diesel engines that produce 4,400 traction horsepower, AC electric traction motors, and 200,000 pounds of starting tractive effort. (See Chapter 5 for an explanation of EPA emissions regulations and Tier 4 locomotives).

Siemens Charger SC-44 at King Street Station, Seattle, WA, Feb 22, 2017. The locomotive powered a WSDOT test train to Vancouver, BC, which left Seattle in push mode and returned the next morning at the head-end. Photo Credit: David Honan.

First EMD F125 locomotive delivered to Southern California Regional Rail Authority's Metrolink, Los Angeles Union Station, July 18, 2016. Photo Credit: Kevin Bleich.

On the U.S. passenger side, the most technologically sophisticated diesel-electric locomotives are the new Tier 4-compliant 4,700 hp EMD F125 Spirit locomotive, 4,400 hp Siemens SC-44 Charger,[24] and 5,400 hp MotivePower MP54AC, and the Tier 3-compliant 4,650 hp MotivePower HSP46.

The EMD F125 Spirit and Siemens SC-44 Charger are 125 mph locomotives. Both locomotives are equipped with a type of regenerative braking that recovers a portion of the electricity generated by dynamic braking and uses it to power the locomotive's head-end power (HEP) system or a combination of HEP and auxiliary energy systems—thus saving energy.[25] Auxiliary energy systems include locomotive blowers and fans, pumps, compressors, cab lighting, heating, and air conditioning, etc.

(Auxiliary and HEP systems are discussed in Sections 2.10 and 2.12, respectively.)

The MotivePower MP54AC, intended for the commuter rail market, uses two 2,700 hp Cummins diesel engines and is the most powerful diesel-electric locomotive on the U.S. market. It is geared for rapid acceleration and a top speed of 110 mph. Each MP54AC locomotive is designed to operate on one or two engines. Two engines provide redundancy, and the locomotive can operate in a fuel-saving single engine mode when pulling shorter, lighter trains during off-peak periods. MotivePower sells this locomotive as a new product but will rebuild an earlier model to the new locomotive's specifications.

Norfolk Southern's fleet of SD70ACe and ES44AC heritage locomotives, Spencer, NC, July 3, 2012. To honor the railroads that became part of Norfolk Southern, these locomotives are painted in the "livery" or paint scheme of these railroads. Photo Credit: © Norfolk Southern Corp.

Evanston, IL, December 2016. Photo by Judie Simpson.

CHAPTER TWO

How Diesel-Electric Locomotive Energy Systems Work

Now let's examine the various systems that make up a modern diesel-electric locomotive, and identify existing energy efficiency technologies, as well as potential future opportunities for efficiency improvement.[26] The focus is primarily on AC locomotives—those with AC inverters and traction motors—but much of the discussion also applies to DC locomotives. The systems we will examine are:

- Locomotive configuration
- Operator's cab
- Electronic control systems
- The Propulsion System
  - Diesel fuel and tank
  - Diesel engine (prime mover)
  - Traction and other alternators/rectifiers
  - Traction inverters
  - AC electric traction motors
- Locomotive trucks and running gear
- Auxiliary energy systems
- Locomotive and train braking systems
- Head-end power
- Locomotive aerodynamics
- Other types of locomotives
  - Genset and slug locomotives
  - Repowering older locomotives

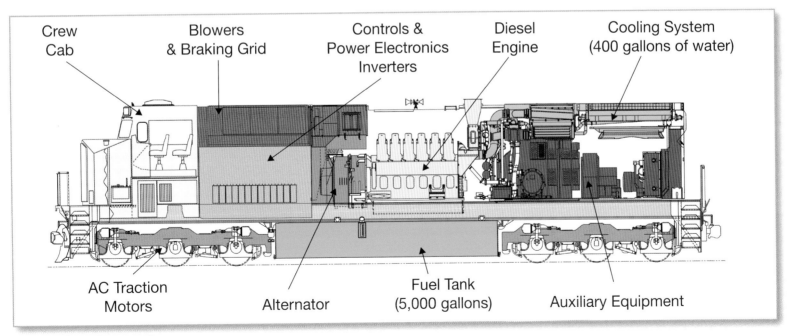

Crew Cab — Blowers & Braking Grid — Controls & Power Electronics Inverters — Diesel Engine — Cooling System (400 gallons of water) — AC Traction Motors — Alternator — Fuel Tank (5,000 gallons) — Auxiliary Equipment

Image Credit: GE Transportation

2.1 Locomotive Configuration

The diagram below shows the configuration or layout of a modern GE diesel-electric locomotive. An EMD diesel-electric locomotive's layout is similar, though dynamic braking grids are located in the rear of the locomotive.

As diesel-electric locomotive manufacturing progressed, locomotive subsystems were organized into modular units, which GE refers to as "cabs." These cabs present another helpful way to visualize the layout of a diesel locomotive. Using cab GE nomenclature, a modern diesel-electric locomotive has five cab areas. These are as follows:

Operator's cab—Located in the front of the locomotive, this is where the engineer and conductor sit to operate the locomotive.

Auxiliary cab—Behind the operator's cab, the auxiliary cab contains various electrical panels, controls, relays, contactors, power supplies, microprocessors, components of the propulsion system (including rectifiers and inverters), and dynamic braking grids in the case of GE locomotives.

Blower cab—This cab contains the traction alternator, one or more secondary alternators (depending on the locomotive), and blowers that cool various pieces of equipment and pressurize the auxiliary cab to prevent the infiltration of road dirt.

Engine cab—As its name implies, the engine cab contains the locomotive's diesel engine and associated piping for fuel, oil, and filtration.

Radiator cab—This is the largest cab, and it houses the diesel engine's radiators and associated cooling fans, the rear traction motor blower, air compressor, and dynamic braking grids in recent model EMD locomotives.

2.2 Operator's Cab

Locomotive operator's cabs have come a long way from the days of steam. In a steam locomotive cab, the engineer operated the throttle and braking systems, but

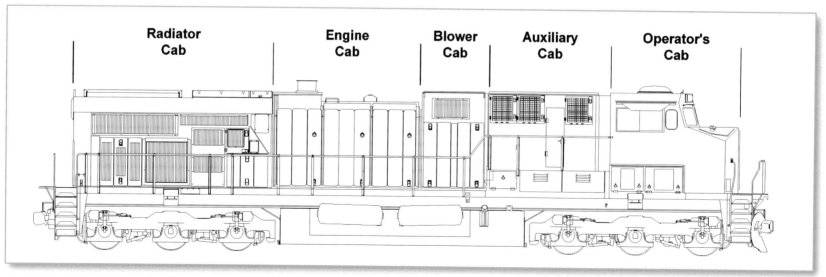

Locomotive "cabs" on the GE 4400DC locomotive Image Credit: GE Transportation.

he and his fireman were also responsible for operating the equivalent of a large boiler room with a steam-driven stoker, feedwater injector, and numerous valves for all kinds of other steam appliances. Careful operation was essential given the potential dangers associated with a 2,500 degrees Fahrenheit (2,500°F) firebox and as much as 300 pounds per square inch (psi) steam. A low water level could produce a boiler explosion that could easily tear the locomotive apart, killing the occupants of the cab and anyone else in the vicinity.

In contrast, modern diesel-electric locomotive cabs have been increasingly designed for the safe operation of a diesel-powered electric power plant coupled to electric traction motors. From the operator's cab of a modern diesel-electric locomotive, the engineer controls and monitors the throttle; dynamic, train, and locomotive brakes; and communication, diagnostic, energy management, signaling, electrical, and safety systems. While early diesel cabs were not as comfortable as the current "North American"

cab design, over time it was understood that a better working environment would improve engineer performance for all tasks, including those associated with energy efficient operation.[27]

Beginning in 1984, locomotive manufacturers equipped their locomotives with microprocessor systems that included diagnostic display panels. These sounded alarms and provided messages identifying components of the locomotive that might not be operating properly. Initially these panels were located behind the engineer, but newer locomotives incorporate diagnostic functions on computer screens placed in the engineer's forward field of view. Digital gauges incorporated in these screens have replaced traditional analog gauges.

Manufacturers also began introducing "wide cabs" or "comfort cabs" to their locomotives in the late 1980s. These cabs were more office-like and comfortable for locomotive crews. Standard AAR (Association of American Railroads) vertical tower control stands were replaced on many locomotives by desk-like horizontal work stands—which in freight locomotives have now fallen out of favor because operators find vertical control stands more comfortable and practical when backing up locomotives. Cabs also became quieter, vibration-isolated, equipped with air conditioning and improved heating systems,

Engineer and chief mechanic Patrick Connors in the functional cab of Buffalo Southern's fully operational RS-18 MLW-ALCO locomotive, built in 1958. Photo by author.

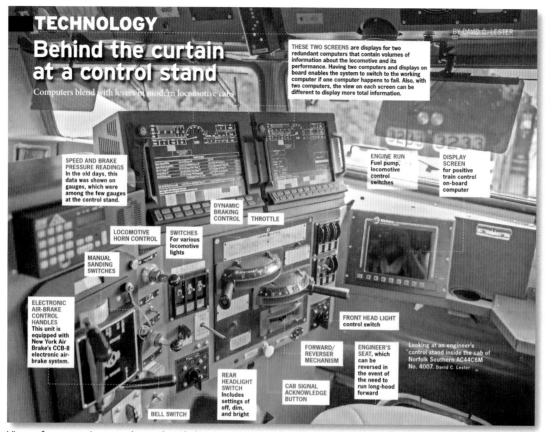

TECHNOLOGY
Behind the curtain at a control stand

BY DAVID C. LESTER

Computers blend with levers in modern locomotive cabs

THESE TWO SCREENS are displays for two redundant computers that contain volumes of information about the locomotive and its performance. Having two computers and displays on board enables the system to switch to the working computer if one computer happens to fail. Also, with two computers, the view on each screen can be different to display more total information.

SPEED AND BRAKE PRESSURE READINGS In the old days, this data was shown on gauges, which were among the few gauges at the control stand.

DYNAMIC BRAKING CONTROL

THROTTLE

ENGINE RUN Fuel pump, locomotive control switches

DISPLAY SCREEN for positive train control on-board computer

LOCOMOTIVE HORN CONTROL

SWITCHES For various locomotive lights

MANUAL SANDING SWITCHES

ELECTRONIC AIR-BRAKE CONTROL HANDLES This unit is equipped with New York Air Brake's CCB-II electronic air-brake system.

FRONT HEAD LIGHT control switch

FORWARD/ REVERSER MECHANISM

ENGINEER'S SEAT, which can be reversed in the event of the need to run long-hood forward

Looking at an engineer's control stand inside the cab of Norfolk Southern AC44C6M No. 4007. David C. Lester

REAR HEADLIGHT SWITCH Includes settings of off, dim, and bright

CAB SIGNAL ACKNOWLEDGE BUTTON

BELL SWITCH

View of operator's control stand with key equipment identified and explained. Image Credit: © 2017, *Trains* magazine, Kalmbach Publishing Co., reprinted with permission. *www.trainsmag.com.*

and had better creature comforts (e.g., refrigerator and improved toilet). Visibility and crashworthiness were also improved.

The comfort cab introduced ergonomic design to the locomotive's "Human Machine Interface" (HMI). *Ergonomics* is the applied science of design that seeks to arrange physical activities so they are as convenient, safe, effective, and efficient as possible. Ergonomic

design was especially important given the increasing number of electronic and computer systems that were placed in the locomotive cab, all vying for the engineer's attention. Among these were displays that showed fuel consumption in real time, thus providing helpful information that could facilitate more energy efficient operation. These additions would have been unmanageable except for designs that sought to consolidate and present them

to the engineer in ways that made them easier to use.

The latest diesel-electric locomotives have a form of "cruise control" more accurately described as an energy management system that works by monitoring a variety of locomotive and route parameters to maximize fuel economy. This technology produces energy efficient operation and would seem to be a plus for engineers. However, this semi-automated running can present its own issues (see Sections 2.3 and 4.2).

Positive Train Control (PTC) is a federally mandated technology that will automatically stop trains before collisions and other types of accidents occur.[28] PTC presents an additional challenge for designers. Its displays must be integrated in such a manner that the HMI promotes its effective use. The HMI is exactly what it sounds like, i.e., the space or place where the human user interacts with machinery or technology.

Heads-up displays, which project data on the locomotive cab front window or windshield, are being developed to allow engineers to monitor key locomotive functions without losing sight of the track. It is not yet clear whether these displays would be helpful or further sources of distraction.[29]

The bottom line for cab design and energy efficiency is how best to organize the cab so that energy management systems and the actions of the engineer work together, optimally, to minimize fuel usage.

2.3 Electronic Control Systems

The electronic control systems in modern diesel-electric locomotives are designed to withstand the physically tough and dirty working environment of locomotive operation. While these systems differ on GE and EMD locomotives, they address the same functions and have been steadily improving with each successive model.

Crew work stations in Brightline Siemens SC-44 Charger passenger locomotive cab. Photo Credit: Siemens Corporation.

Locomotive electronic control systems have three functional parts: (1) an operator interface including the cab control stand—either analog (gauges, switches, etc.) or digital (computer displays and touch screens, etc.), (2) a control system that includes a computer or multiple computers, relays, panels, electrical switchgear, power converters, etc., to control locomotive functions and engine support functions, and (3) an engine controller, which on older locomotives would have been a mechanical governor, but on newer locomotives is an electronic fuel injection controller/computer.

Alternatives to annoying audio "alerters" are also under investigation. Alerters are designed to keep engineers awake and attentive. They are used in all U.S. long-haul passenger locomotives and most freight locomotives. Depending on speed, they sound off every 30-120 seconds unless the engineer operates specific locomotive controls or pre-emptively hits the alerter button before the alarm sounds. Optical alerter systems, which use cameras facing the operator and software to examine operator eye movements, may be viable in the future.

The Federal Railroad Administration has sponsored research on the "Next Generation Locomotive Cab" for U.S. freight locomotives. This research has used the vertical side-mounted Association of American Railroads' AAR-105 control stand as a starting point, simplifying it and rearranging monitors and other controls to improve ergonomics and human factors.[30]

GE and EMD have their own names for various systems. For example, the GE locomotive operator interface consists of Smart Displays, while EMD locomotives have FIRE Displays. As part of their locomotive control systems, both manufacturers have AC traction controllers that control their locomotive's traction alternator, rectifiers, inverters, AC traction motors, and dynamic braking. Locomotive control systems rely on input/output panels that bring together inputs from sensors scattered throughout the locomotive and send output signals to various pieces of equipment to operate them.

Locomotive Energy Management Programs

Diesel-electric locomotive computer systems use a number of computer programs, i.e., types of software, designed to enhance energy efficiency. These tend to be "adders," purchased as upgrades to base-level freight locomotives, or after-market products. While freight railroads use these systems extensively, as of 2017 Amtrak did not use them—preferring that its engineers remain in full control of locomotives at all times. (Locomotive energy management systems are also discussed in Section 4.2.)

Among locomotive energy management systems are the following, organized by manufacturer:

General Electric Locomotive Software

Trip Optimizer™—To optimize energy efficiency, the *Trip Optimizer* program uses an energy conservation algorithm that develops an individual plan for each train based on train parameters (e.g., weight, length, and horsepower) and route parameters (e.g., location, length of trip, gradients, and track curvature). It then operates the locomotive as a kind of smart "cruise control," automatically controlling the locomotive throttle, dynamic brakes, and distributed power functions to reduce fuel consumption and arrive on time. GE claims that its Trip Optimizer produces fuel savings of 3% to 17%.[31]

Canadian Pacific tested it and reported that it achieved 6% to 10% savings,[32] while its bulk commodity trains achieved up to 20%. Originally designed only for GE locomotives, the Trip Optimizer is now available after-market for EMD locomotives.

Trip Optimizer was preceded by another GE locomotive energy management

Progress Rail EMD SD70ACe-T4 cab and vertical control stand. Note placement of computer screens and improved sight lines. Photo Credit: Chris Bandel.

program called *Trip Advisor*. This earlier program provided prompts to engineers to encourage them to make operational changes to improve efficiency.

Trip Optimizer is part of a "suite" of GE railroad software programs. Branded "RailConnect 360," these programs include Power Advisor (to assign the most efficient locomotives to a given train), Movement Planner (to improve the efficiency of dispatching in a rail network), and Locotrol, which is described next.[33]

Locotrol®—Pioneered by the Southern Railway over 50 years ago, Locotrol is GE's control system for distributed power. Locotrol allows the lead locomotive to control additional locomotives in the middle or at the rear of the train. It functions wirelessly, producing the benefits of distributed power operation including energy efficiency.[34] (Distributed power is discussed in Section 4.5.)

Progress Rail EMD Locomotive Software

SmartConsist™ Fuel Management System —Once the engineer selects a throttle setting, this program automatically adjusts the throttles in all of the locomotives in the train to achieve the desired total horsepower while maximizing energy efficiency. Similar to GE's Locotrol, *SmartConsist* is available as an option for new EMD locomotives and as a retrofit for in-service

EMD locomotives. EMD reports fuel savings of 1% to 3%, for a quick payback. Note that a *consist* is a group of locomotives.

IntelliTrain®—A cell phone and GPS monitoring system that permits remote analysis of locomotive operation and provides maintenance recommendations.

Progress Rail advertisement extolling the energy efficiency of EMD locomotives and LEADER with a "Less Is More" energy conservation slogan (*Railway Age*, July 2017). Image Credit: Progress Rail, A Caterpillar Company.

New York Air Brake Locomotive Software

LEADER® (Locomotive Engineer Assist/ Display & Event Recorder)—Of the two primary locomotive energy management systems, New York Air Brake's LEADER was first on the market—initially as an operator prompt program. Functionally similar to GE's Trip Optimizer, LEADER is described as a "real-time train handling system" because of its ability to "look-ahead" in real time and continually update its operating plan for improved energy efficiency and on-time arrival. NYAB's expertise in locomotive simulators and train handling provided a basis for LEADER, which primarily operates on EMD locomotives, but more recently has become available for retrofit on GE locomotives.

NYAB's LEADER can operate in a "Driver Assist" mode, prompting the engineer to make throttle and brake adjustments to maximize fuel efficiency, or in "Auto Control" mode to achieve additional savings by taking over throttle and dynamic braking control from the engineer.[35] Additionally, the program is capable of providing *asynchronous distributed power* control, i.e., different throttle settings for locomotives located in different sections of a long train.

27

Norfolk Southern, which has used LEADER extensively and played a key role in adapting LEADER to GE locomotives, reports that its LEADER-equipped locomotives show an average 5% to 7% improvement in fuel economy.[36] NYAB reports fuel savings of 6% to 17%.[37] LEADER also functions as an event recorder, which is described below.

Wi-Tronix Software

EcoRun™ [38]—This program is described as a "driver advisory system" that helps engineers select optimal throttle settings to improve fuel efficiency.It also makes efficient idling and coasting recommendations and warns against inefficient stretch braking. Claimed energy savings are greater than 10%. Wi-Tronix, an after-market product manufacturer whose Eco-Run has less market penetration than Trip Optimizer and LEADER, also offers an event recorder[39] and telemetry program that remotely monitors locomotive operations and energy consumption. (Note that *stretch braking* occurs when train brakes are applied while the throttle is still engaged. This stretches the train to avoid railcar bunching. While there are circumstances when stretch braking makes sense in train operation, as a general practice during normal running it does not—and, in those circumstances, it wastes fuel.)

Event Recorder

A diesel-electric locomotive's event recorder serves the same function as an airliner's black box and is critically important when accidents occur. An event recorder is defined by federal regulations as a "device, designed to resist tampering, that monitors and records locomotive operating data over the most recent 48 or more hours of operation." These devices can also serve an energy management function because they collect locomotive throttle and fuel use data. (Locomotive event recorders are also discussed in Section 4.2.)

The Federal Railroad Administration requires event recorders on all locomotives that operate at 30 mph or more. For locomotives ordered on or after October 1, 2006, or placed in service on

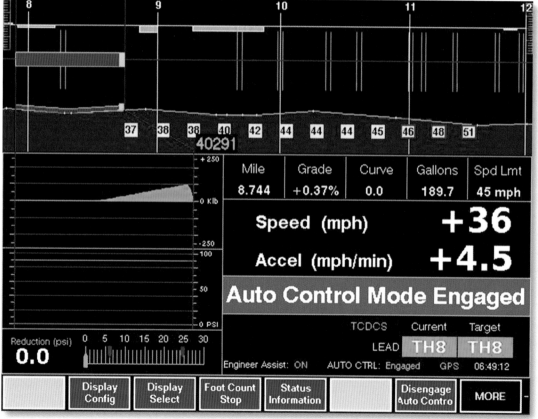

LEADER's "Auto Control Mode Engaged" screen shows locomotive speed, acceleration, location, gradient, track curvature, throttle notch, and fuel consumption (in this case 189.7 gallons per hour). Image Credit: New York Air Brake.

or after October 1, 2009, event recorders must record these and other parameters:

- Speed
- Direction of motion
- Time
- Throttle position
- Braking systems status
- Cab signaling
- EOT (end-of-train) device status
- Headlight and auxiliary light operation
- Tractive effort
- Energy management (Trip Optimizer or LEADER) status[40]

This information will eventually be integrated in a single Positive Train Control (PTC) event recorder.

Positive Train Control

PTC is a processor/communication-based train control system that provides engineers advance warning of track conditions so that they can prevent collisions, derailment, or other accidents due to excessive speed, conflicting train movements, or movement through improperly lined main-line switches.[41] PTC automatically slows or stops a train if an engineer does not take action in time. Most Class I railroad and Amtrak diesel-electric locomotives must be equipped with Positive Train Control systems by December 31, 2018, unless the U.S. Congress extends the deadline again. The industry price tag for PTC is estimated to be in excess of $10 billion once fully deployed.[42] The Association of American Railroads (AAR) has warned that PTC may reduce capacity and thus make rail operation less efficient due to its complexity and untested nature, and, therefore, should not be rushed.[43]

Electronic Control Systems Energy Efficiency Opportunities. The advent of microprocessor diesel engine control systems in locomotives, beginning in the 1980s with GE's Dash 8 locomotive, resulted in greater efficiency. Additional efficiency gains are possible as computer control systems for locomotive energy-related subsystems improve with each new locomotive model.

GE and EMD have steadily improved their locomotive energy management systems. Further improvements are possible, however, by modifying Trip Optimizer and LEADER to be more user-friendly to engineers who want and need to be full participants and not simply passive observers of the technology.

EMD GP38-2 locomotive refueling at the BNSF Cicero Yard, Chicago, IL, November 2015. Photo by author.

Event recorders can also be improved by enhancing their ability to serve positively as teaching tools for engineer training. Obviously, the effective integration of PTC should be a priority.

2.4 Diesel Fuel and Fuel Tank

Providing diesel fuel to a diesel-electric locomotive's diesel engine is a 3,000 to 5,300-gallon fuel tank that is bolted or welded to the locomotive's frame, though some passenger locomotives have integral fuel tanks to reduce overall weight.[44] At 7 pounds a gallon, 5,000 gallons of diesel fuel weigh 35,000 pounds or 17.5 tons. Thus, nearly 10% of the weight of a 200-ton diesel-electric locomotive could be its fuel. A GE ES44AC or EMD SD70ACe locomotive will consume about 200-210 gallons of diesel fuel an hour at full power (Notch 8). At that rate, a 5,000-gallon fuel tank would last about 24 hours. If a diesel-electric locomotive is well-utilized, it can readily consume 300,000 gallons of diesel fuel a year—equal to 60 full fuel tanks.

Diesel-electric locomotives are generally refueled at terminals and main-line fuel pads. Long trains, with multiple locomotives located in the rear and/or middle as well as head-end of the train in a distributed power configuration, may use tanker trucks so that all locomotives can be refueled simultaneously.[45] Locomotive fuel tanks are refilled at varying rates, de-pending on method, e.g., 200 gallons per minute if by tanker truck and up to 600 gallons per minute at refueling stations.[46]

Petroleum-based diesel fuel is made from crude oil.[47] The crude is heated to over 1,000°F, a temperature at which it evaporates into a variety of gases that rise in a distilling or *fractionating* column. Various diesel hydrocarbon molecules condense at different temperatures (360°F-680°F) as they cool off while rising in the tower. This condensate is collected in trays placed in the tower. A process called *hydrodesulfurization* is used to remove most of the sulfur from the fuel, creating ultra-low sulfur diesel fuel that locomotives are required to use.

The energy content of diesel fuel is given in two ways—the Higher Heating Value (HHV) and the Lower Heating Value (LHV).[48] The HHV is the total energy content of the fuel. It is the amount of energy that is released when the fuel is completely combusted. The LHV is the "net

Fuel and air reserve tanks on a CSX GE AC6000CW locomotive. These 6,000 hp locomotives were classed CW60AC by CSX. Cumberland, MD, June 2014. Photo by author.

energy content" of the fuel and is equal to the HHV minus the amount of energy expended vaporizing water in the combustion products. Energy calculations involving internal combustion engines generally use the LHV because the latent heat of vaporization of water is not available for use by these engines. Diesel fuel's HHV is 138,000 British Thermal Units (BTUs) per gallon; its LHV is 128,000 BTUs per gallon. One BTU equals the amount of heat it takes to raise the temperature of a pound of water 1°F.

Diesel fuel gets cloudy and gels as its temperature drops. Naturally occurring waxes in the fuel begin appearing as small crystals when the fuel reaches its *cloud point* temperature. The *pour point* temperature is reached when the fuel has gelled sufficiently that it will no longer pour. If waxes are allowed to precipitate in the fuel, fuel filters and strainers can become plugged and the engine will become starved for fuel.

"Waxing" can be suppressed by using fuel additives known as *wax crystal modifiers*. Railroads have operating practices, codified as operating rules, which call for the injection of anti-gel fuel additives into storage tanks to prevent waxing. A railroad may also choose to specify from its supplier diesel fuel that has lower cloud points when the fuel is intended for winter use. For example, Union Pacific, has

specified that its winter diesel fuel must have cloud points between 5°F and 12°F, depending on the expected winter temperatures of its various routes.[49]

To prevent diesel fuel from cooling below its cloud and pour points, it is heated in two ways. First, locomotive diesel engines are designed to receive excess fuel from the fuel tank in order to ensure an adequate supply to the fuel injectors and to cool and lubricate them. This excess fuel is returned to the fuel tank at approximately 90°F. Second, the fuel line from the fuel tank to the engine contains a thermostatically controlled fuel preheater that, when outside temperatures drop, uses hot engine coolant water to warm the fuel as needed. Typically, this liquid-to-liquid heat exchanger is located on the floor of the engine cab.

Diesel fuel has a shelf life and degrades in long-term storage. Warm temperature and the presence of water and contaminants can accelerate oxidative degradation, leading to non-combustible sediments and gums forming in the fuel. Additives can stabilize diesel fuel and prolong its useful life.

The price of locomotive diesel fuel varies year to year (actually day to day on the spot market) and plays a role in energy conservation. Presently, fuel prices are historically low in current dollars and may remain relatively low for years as a result of U.S. energy policy and increasing U.S. "shale oil" supplies. Shale oil is recovered deep underground using hydraulic fracturing drilling methods (see Section 6.4).

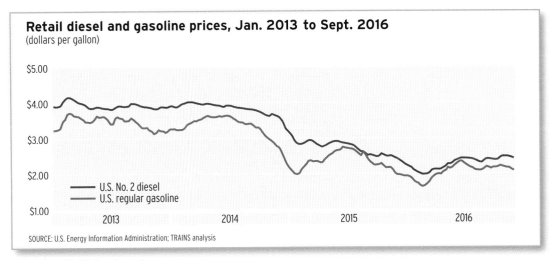

Retail diesel and gasoline prices, Jan. 2013 to Sept. 2016
(dollars per gallon)

— U.S. No. 2 diesel
— U.S. regular gasoline

SOURCE: U.S. Energy Information Administration; TRAINS analysis

Image Credit: *Trains* magazine.

Diesel Fuel Tanks vs. Coal Tenders—Which Contain More Energy?

Let's compare the energy content of a diesel-electric locomotive fuel tank filled with 5,000 gallons of diesel fuel to the energy content of a steam locomotive tender carrying 26 tons of coal.

This comparison can be calculated as follows using the lower heating value of diesel fuel:

Energy content of 1 gallon of diesel fuel = 128,000 BTUs
128,000 BTUs/gallon x 5,000 gallons = 640,000,000 BTUs

Energy Content of 1 ton of coal = 28,000,000 BTUs
28,000,000 BTUs/ton x 26 tons = 728,000,000 BTUs

640,000,000 BTUs/728,000,000 BTUs = 0.879 or 87.9%

Thus, we see that the fuel tank of a modern diesel locomotive carries nearly as much energy as the much larger coal bin in the Y6's tender. How can this be? It's possible because oil has a much higher energy density than coal. However, it is the much greater efficiency of the diesel-electric locomotive that explains why it can operate at full power for nearly 24 hours before refueling, while the Y6 operating at full output would burn through its giant tender's coal supply in a matter of hours.

Norfolk & Western Y6 articulated steam locomotive with tender, which holds 26 tons of coal as well as 95 tons (23,000 gallons) of water. Photo Credit: © Norfolk Southern Corp.

Diesel Fuel Energy Efficiency Opportunities.
While diesel fuel prices were above $3.00 per gallon for much of the 2011-2015 period,[50] Union Pacific reported that its price of diesel fuel was just $1.48 per gallon in 2016.[51] This low fuel price was presumably still high enough for UP to keep energy efficiency and conservation in mind, given the nearly 1 billion gallons of fuel the railroad's locomotives consumed that year.[52] That said, as a general rule, interest in energy conservation is higher when energy prices are higher.

2.5 Diesel Engine or "Prime Mover"

Most diesel engines used in diesel-electric locomotives are V-12 or V-16 designs. These big engines weigh 15 tons each and are designed to be as durable as possible. The new V-12 engine used by EMD's newest locomotive, the SD70ACe-Tier 4, has a total displacement of 12,120 cubic inches—or over 1,000 cubic inches per cylinder. Relative to smaller internal combustion engines, these big engines operate slowly with a maximum rotational speed of only about 1,000 rpm. As previously explained, diesel engines do not use spark plugs to ignite their fuel. They have high compression ratios, enabling them to compress combustion air to a point where its temperature is sufficient to ignite diesel fuel when it is atomized (sprayed) into the cylinder.

The energy efficiency of the prime mover can be understood as a function of these four types of engine efficiency:

- **Combustion**—The efficiency of turning chemical energy in diesel fuel into heat energy (typically well over 99%)
- **Thermodynamic**—The efficiency of converting combustion heat into pressure that pushes the pistons
- **Gas exchange** – The efficiency of the engine pumping in clean air (with oxygen to support combustion) and exhausting out combustion gases

- **Mechanical**—The efficiency of the engine's moving parts (e.g., gears, crank and cam shafts, and valve mechanisms) minus engine-related parasitic loads (e.g., fuel, water, and oil pumps)

Where, Combustion eff. x Thermodynamic eff. x Gas Exchange eff. x Mechanical eff. = Engine Efficiency

GE Locomotive Assembly Hall, Erie, PA, locomotive shop, March 2006. Photo Credit: Sean Graham-White.

Installation of a GEVO-12 engine in an ES44AC locomotive. Photo Credit: Greg McDonnell.

The prime movers of modern diesel-electric road locomotives typically produce 4,000-4,400 *traction horsepower*—the amount of horsepower the engine delivers to the traction alternator for the purpose of generating electricity for the traction motors that propel the locomotive. The diesel engine itself actually produces 100 to 300 horsepower more than that. The additional power is used to generate electricity to power *auxiliaries* or *parasitic loads* within the locomotive, i.e., everything else requiring power. The total amount of horsepower that the prime mover produces is called *brake horsepower* and can be thought of as horsepower at the engine crankshaft.

Prime movers in locomotives may be either four—or two—cycle engines (previously discussed in Section 1.3). GE locomotives use four-cycle engines, while EMD locomotives have historically used two-cycle engines. The newest EMD locomotives, its Tier 4 freight and passenger locomotives, use new four-cycle diesel engines to achieve efficient operation while complying with more restrictive emissions standards.

Engine and Locomotive Operation

Starting the prime mover involves multiple steps. First, electricity from the locomotive's massive collection of DC batteries is converted to AC power by inverters and used to start a sequence of fuel and lubrication oil pumps. As a general rule, GE locomotive engines have one lube oil pump, while EMD engines have three—two pumps for the engine block (one delivering oil to the block's moving parts and the other scavenging or pulling oil out of the crankcase) and one pump that provides oil to the turbocharger. Then, the engine is rotated or cranked for starting. EMD uses a pair of cranking starter motors that turn a large ring gear attached to the engine's flywheel, while GE uses a "cranking inverter" to briefly operate the traction alternator as a variable speed AC motor that rotates the crankshaft and starts the diesel engine.

Once the prime mover is started, the engineer can operate the locomotive by first moving the "reverser" lever on the control stand to a forward or reverse position, and then moving the throttle lever from idle through a series of throttle settings, or "notches," numbered 1 to 8, where 8 is full power.

The throttle is designed so that each higher notch provides a discrete amount of additional power. However, the operation of the locomotive's traction control system causes the horsepower output in each throttle notch to vary as the system seeks maximum adhesion, with the notch selection providing the *maximum* available horsepower for that notch. Although the power levels associated with different notches on GE and EMD locomotives are slightly different, they provide roughly 5%, 15%, 25%, 35%, 50%, 67%, 85% and 100% of full power for Notches 1–8, respectively.

When accelerating the locomotive and train, the engineer must carefully add increments of horsepower to avoid wheel slippage or sudden excessive traction that could snap a coupler (*knuckle*) or its shank. Knuckles are rated for 400,000 pounds, the amount of tractive effort just two AC locomotives can generate on starting. When starting on level track, a gentle throttle can effectively pull out slack or stretch the train one car at a time—thus, minimizing both the amount of tractive force needed to start a train and the risk of snapping a knuckle.

To reduce power, the engineer moves the throttle in the reverse direction, i.e., to lower notches or throttle positions. Once at idle, the engineer may apply increments of dynamic braking by using the same throttle lever on a desktop console, or a separate dynamic braking lever on a vertical control stand.

Engine Control Unit

When microprocessor controls were applied to locomotive engine operation, the engine control unit (ECU) replaced the mechanical governor. The ECU only operates the diesel engine. It translates commands from GE HMI Smart or EMD FIRE displays or the control stand into appropriate throttle settings, engine speeds, horsepower, and levels of fuel consumption. Engine speed and output are controlled by processors that adjust fuel delivery and injection timing based on inputs from numerous sensors. For example, the ECU on a GE Tier 3 locomotive is said to receive diesel engine input signals from nearly 50 sensors. The number increased substantially on Tier 4 locomotives.

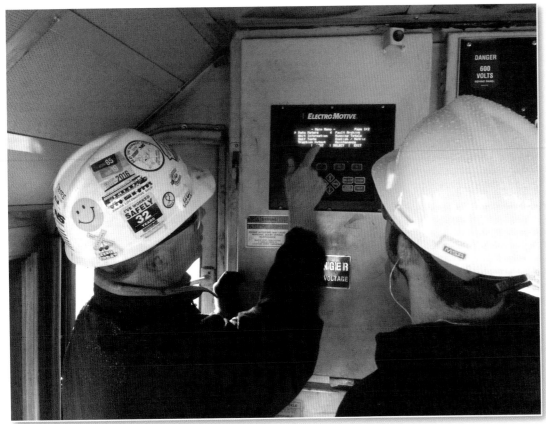

Norfolk Southern's Juniata Locomotive Shop Plant Superintendent Donald R. Faulkner demonstrates diagnostic display panel to Penn State Altoona railroad engineering student. Photo Credit: Bryan Schlake.

ECU sensors on modern diesel-electric locomotives measure the following:

- **Speed** of engine crankshaft and turbocharger(s)
- **Temperature** of the diesel fuel, lubrication oil, ambient and combustion air, the exhaust manifold, cooling water, turbochargers, and various radiators and other heat exchangers
- **Pressure** of engine manifold, fuel, lubrication oil, cooling water, air filter, and crankcase, along with atmospheric pressure

Diesel engines have many other important systems. These include fuel injection, combustion air, lubrication oil, and cooling water systems.

Fuel Injection System

A fuel transfer pump delivers diesel fuel from the tank to the engine through low-pressure fuel lines. The fuel is then further pressurized by an additional pump or pumps before being injected into the engine's cylinders by fuel injectors. The injectors are electronically controlled valves that push just the right amount of fuel into each cylinder at the top of its stroke. Injection timing is computer-controlled.

Fuel injectors are designed to atomize and distribute fuel in the cylinders. The fuel's burning pattern, which can be manipulated by various design considerations, affects the thermodynamic efficiency of the engine, i.e., how

EMD 4,000 hp, 16-cylinder, 710 diesel engine with top deck cover removed. Fuel manifold (pipe *on left*) supplies diesel fuel to electronic fuel injectors via stainless steel braided fuel lines that have replaced the original copper tubing. Camshaft, *along the top*, controls rocker arm assemblies that open and close exhaust valves and compress spring-loaded injectors in this two-stroke prime mover being refurbished at Norfolk Southern's Juniata Locomotive Shop in Altoona, PA. Photo by author.

efficiently energy of combustion is transferred to the pistons. To reduce emissions while maintaining efficiency, GE and EMD use high-pressure common rail fuel injection systems in their Tier 4 locomotives. (See Section 5.2.)

Combustion Air System

On modern diesel-electric locomotives, the prime mover has one or more turbochargers. A turbocharger is a waste heat recovery device that uses the engine's red-hot, high-velocity exhaust stream to turn a turbine. The shaft of the turbine turns a compressor that compresses combustion air so more of it (and thus more oxygen) can be forced into the engine's cylinders. This increases power output. It also increases the gas exchange efficiency of the engine by using waste energy to help the engine "breathe" more easily.

Turbochargers on four-cycle diesel engines are free-running and don't provide effective boost and energy recovery until the engine is operating in higher notches, e.g., 6, 7, or 8, where exhaust waste energy is higher. At lower notches, the four-cycle engine's intake stroke does all or most of the work supplying combustion air to the cylinders. In contrast, two-cycle engines do not breathe on their own and require that combustion air be blown into their cylinders at all throttle settings. This is accomplished by a turbocharger that is mechanically connected to the engine's crankshaft. In lower notches, the engine's crankshaft drives the turbocharger through gears and an overriding clutch. In higher notches, when there is sufficient exhaust gas energy to power the turbocharger, the clutch disengages the turbocharger from the crankshaft, and the turbocharger becomes free-running like turbochargers on four-cycle engines.

Turbocharged combustion air is not only pressurized, but it's hot and must be cooled in order to increase its density and ability to deliver more oxygen to the cylinders. Combustion air is cooled by heat exchangers called *intercoolers*, *aftercoolers*, or just *coolers*. Typically, there are two air-to-water heat exchangers, each serving a bank of cylinders. These intercoolers (GE terminology) or aftercoolers (EMD terminology) cool the combustion air by transferring its heat to the cooling water that circulates in either the locomotive engine's cooling water system or a separate cooling loop with its own radiator—see "split cooling" later in this section. The radiators reject the heat into the atmosphere.

In order to decrease emissions, GE Tier 2 and 3 Evolution series locomotives further reduce the temperature of turbocharged combustion air with fan-cooled air-to-air heat exchangers.[53] These heat exchangers are located downstream of the locomotive's air-to-water intercoolers.

Introducing relatively cool combustion air into the cylinders is so important for efficiency and emissions that the outer casings and compressor impeller blades of locomotive turbochargers may themselves be water-cooled.[54]

While a single turbocharger extracts useful energy from the engine's exhaust stream, by no means can it extract and use all of it; a lot of energy is still lost through the locomotive's exhaust stack. This realization has produced compound turbocharger schemes where multiple turbochargers are placed in a cascading series to extract more energy from the exhaust waste stream. To meet Tier 4 emissions

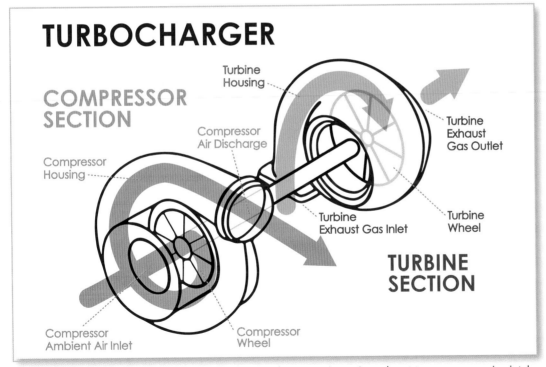

TURBOCHARGER

COMPRESSOR SECTION

Turbine Housing

Compressor Air Discharge

Compressor Housing

Turbine Exhaust Gas Outlet

Compressor Ambient Air Inlet

Compressor Wheel

Turbine Exhaust Gas Inlet

Turbine Wheel

TURBINE SECTION

Concept drawing of turbocharger showing how it uses the energy in engine exhaust to compress engine intake combustion air. The common shaft connecting the turbine and compressor portions of the turbocharger can be seen. On locomotive diesel engines, turbochargers have two combustion air outlets, one for each bank of cylinders (*see next page*). Image Credit: Anne Johnston Fera.

Turbochargers Force Combustion Air into Engine Cylinders

Contrasting combustion air systems on older EMD 645 and 710 engines. The diesel engine in the middle is a "normally aspirated" or "roots blown" prime mover whose combustion air is supplied by a pair of mechanically driven superchargers, each serving a bank of cylinders. These superchargers use a pair of spinning screws which mesh to compress and pump air – hence their oval shape. In contrast, the newer prime movers on either side are supplied with combustion air from large turbochargers whose dual outlets can be seen supplying combustion air to each engine cylinder bank. The circular openings in the middle of these turbochargers reveal inlet air vanes. Over the years, EMD 645 and 710 diesel engines have been railroad industry workhorses and there have been more than a dozen models of these engines. Photos by author.

A refurbished 16-cylinder GE 7FDL prime mover at Norfolk Southern's Juniata Locomotive Shop. The engine's single large turbocharger, seen here prominently on the top front of the engine, has two outlets—each providing hot combustion air to intercoolers serving a bank of cylinders.

Intercoolers Cool Combustion Air to Meet Emissions Standards

GE 7FDL prime mover intercooler. In this photograph, hot turbocharged combustion air enters the intercooler on its left side. Water-filled cooling coils inside the intercooler reduce the temperature of the combustion air which exits through the opening shown on its right side before being directed to all cylinders in the cylinder bank. Photo by author.

In order to meet emissions standards, GE Tier 2 and 3 locomotives provided additional combustion air cooling with air-to-air heat exchangers whose heat rejection vents were mounted in the cooling cab roof between the prime mover's exhaust stack and its main radiators. Small fans were used to pull air through these heat exchangers. Photos by Paul Withers (Diesel Era).

Union Pacific Tier 2 ES44AC locomotive quickly approaches Depew, NY, Amtrak station, on cold day in March 2014. Photo by author.

standards without sacrificing energy efficiency, both GE and EMD adopted multi-stage turbocharging systems. Sequential turbocharging systems—with low- and high-pressure turbochargers—can improve turbocharger boost and effectiveness at lower engine speeds.

Lubrication Oil System

Diesel engines in modern diesel-electric locomotives are lubricated by 400 gallons of lube oil. This oil lubricates the babbitt-like brass and bronze crankshaft and connecting rod bearings—simultaneously removing heat from these bearings so that they can survive the harsh engine environment. Lube oil lubricates the piston walls and rings so that pistons move with minimum friction inside the cylinders and heat is transferred to the engine block to facilitate cooling. A spray of oil to the bottom surfaces of pistons helps to cool them. Lube oil also lubricates and cools turbocharger bearings. In addition, it helps seal cylinder combustion chambers (by forming a film between the piston rings and the cylinder wall), cushion engine parts against shocks, remove dirt, prevent corrosion, and quiet the engine.

Engine lubrication oil is circulated by an oil pump, and the oil is cooled by an oil cooler, which is a liquid-to-liquid heat exchanger that transfers heat from the oil to engine cooling water. This heat is rejected into the atmosphere by the engine's primary radiators.

Cooling Water System

The diesel engine block and various heat exchangers serving combustion air and lube oil (as well as exhaust gas recirculation in Tier 4 locomotives) are cooled by 300-400 gallons of *cooling water*. The heat this water absorbs from all cooling loads is rejected into the atmosphere by large radiators. These radiators are located in massive wing-like structures evident on the rear end of modern GE and EMD locomotives.

Radiators. The "wings" are the result of the manufacturers' search for more radiator heat transfer area as cooling

demands increased. Traditionally, locomotives had two large radiators mounted horizontally adjacent to one another in the upper portion of a straight-sided radiator cab. Extending the length of these radiator banks could have provided more cooling surface, but only if the locomotive platform was lengthened—an approach neither the railroads nor builders liked. A better solution was to widen the top portion of the radiator cab so wider radiators could be enclosed. An enhancement to this strategy was to install the two radiators in a slight V configuration. That allowed for

CSX GE AC6000CW locomotive showing off its radiator cab. The air intake vents for the locomotive's massive radiators are just below the wing structure. Cumberland, MD, June 2014. Photo by author.

Top-view of GE Tier 4 locomotive radiator cab showing radiator outlet grills. The first set of vents exhaust heat from the locomotive's intercooler radiators; the second set of vents exhaust heat from the primary radiators. The large size of these radiators is indicative of the Tier 4 locomotive's increased cooling load. Photo Credit: Paul Withers (Diesel Era).

radiators of even greater width and therefore increased cross-section. The "flaring out" of the radiator enclosure began with EMD's SD45 locomotive.

Radiator Fans. Modern diesel-electric locomotives have one, two, or more radiator cooling fans. These fans push or pull air across radiator coils to increase the transfer of heat from the water flowing inside the coils to the air moving across them. Waste heat is rejected at the top of the radiator cab. On GE locomotives, the fans are located below the radiators and push air upward through them. On EMD locomotives, the fans are generally

located on the locomotive roof and they pull air through the radiators.

Radiator fans initially operated on/off and then with two or three discrete speeds. Now, with inverter-controlled AC motors, cooling fans operate at variable speeds. Variable speed operation has improved fan and cooling system efficiency. Note that earlier EMD locomotives used direct-drive cooling fans that were physically connected to locomotive diesel engine crankshafts through a series of gears. In these locomotives, cooling fan speed varied in direct proportion to engine speed—an arrangement which made sense.

Shutters. Some locomotives use shutters, a type of rotating damper like a venetian blind, as part of their cooling systems. These shutters are either

Side-mounted shutters and top-mounted radiator fans on Union Pacific EMD SD70M locomotive. Altoona, PA, February 2017. Photo by author.

opened or closed. They are used to help control air flow across the radiators. All EMD locomotives have shutters, and they are generally located on the side of the locomotive radiator cab just below the top of the wing. As such, they are on the inlet side of the radiator. GE has used shutters sparingly. They appeared on certain 1993 GE Dash 8 locomotives which were given split-cooling and electronic fuel injection. GE also used shutters in conjunction with the air-to-air heat exchangers used to cool combustion air in its Tier 2 and Tier 3 locomotives. Those shutters were located on the inlet side of the heat exchangers.

Dry vs. Wet Radiators. There are many cooling water system variants, and the manufacturers have favored different designs. For example, GE locomotives prior to its "Evolution Series" used a *dry radiator system*. These drained cooling water from radiators when cooling was not required. Now all U.S. locomotive builders use a *wet radiator system* that operates in a manner similar to car and truck engine-block-cooling systems—where radiators always have coolant in them.

Water Temperature. Locomotive cooling water systems are designed to keep engine water temperature in the 185°F-195°F range regardless of engine output, other cooling loads served by the cooling system, and ambient (outdoor) temperature conditions. This engine temperature range is a compromise be-

tween competing interests. On the one hand, engine cooling is essential to keep the engine block gaskets cool enough to avoid abrupt failure and reciprocating metal parts cool enough to avoid being weakened by thermal stress. On the other hand, it is thermodynamically desirable to operate with cooling water as hot as possible in order to minimize energy losses to the engine block. These competing interests have led to an engine water temperature compromise – i.e., cool enough to prevent damage but hot enough to be efficient.

Water as Coolant Choice. It may be surprising that diesel-electric locomotives use water (with rust inhibitors) as the coolant in engine cooling systems instead of an antifreeze (glycol) and water mix like that used in a car or truck. Water is used for a variety of reasons, including:

- **Heat transfer efficiency**—Water transfers heat more efficiently than antifreeze.
- **Cooling system size and costs**— Great efficiency of water-based cooling systems permits smaller radiators, which cost less and are desirable when space is at a premium.
- **Coolant purchase and disposal costs**—Water is less expensive than antifreeze.
- **Protection of engine bearings**— Antifreeze can damage engine bear-

ings in older EMD locomotives after leaking into engine lube oil.
- **No confusion about freeze protection**—When antifreeze is used in only some locomotives but not all locomotives, freeze protection in the locomotives using antifreeze could be lost without warning if maintenance staff mistakenly add water instead of antifreeze when coolant is low. (The SD90MAC-H locomotive uses antifreeze.)
- **Environmental protection**—If coolant leaks, no harm is done if it is water.

Of course, the downside of using water instead of antifreeze is that water freezes at a higher temperature, and freeze-ups can cause engine block and pipe cracking. Traditionally, railroads dealt with potential freeze-up problems by idling locomotives all winter just as they did with steam locomotives. Fortunately, idle reduction technologies now keep coolant water from freezing while using less fuel and producing less pollution. That said, there are still energy costs associated with these technologies.

As a last line of defense against freezeups, inactive locomotives have their cooling water drained. If that isn't done, and idle reduction technologies are inoperative, locomotive coolant systems have emergency drain valves or plugs that pop open and drain the water when water temperature drops below a prescribed level. This saves the engine from freeze damage but at the same time disables it.

Locomotive engine cooling systems must operate effectively—with minimal power and efficiency loss—in extreme, unforgiving conditions. These cooling systems are put to the test when ambient temperatures soar past 100 degrees. Engine cooling also becomes a challenge when locomotives in multiple-locomotive consists operate in high throttle settings in long tunnels.

Split-Cooling. The GE Dash 8 (manufactured 1983-1994) was the first diesel-electric locomotive with a split-cooling system—called *split* because two or more cooling loops with additional radiators and/or heat exchangers are used to cool various engine components. Split-cooling makes it possible to provide lower temperature cooling water to combustion air heat exchangers (again, GE *intercoolers*; EMD *aftercoolers*) than is needed to cool the engine block. In Tier 4 locomotives, elaborate split-cooling systems also provide appropriate temperature water to exhaust gas recirculation coolers.

Locomotive manufacturers have used a variety of different split-cooling configurations. The GE Dash 9 and EMD SD70MAC split-cooling configurations shown here are just examples.

GE DASH 9 Locomotive Split-Cooling System

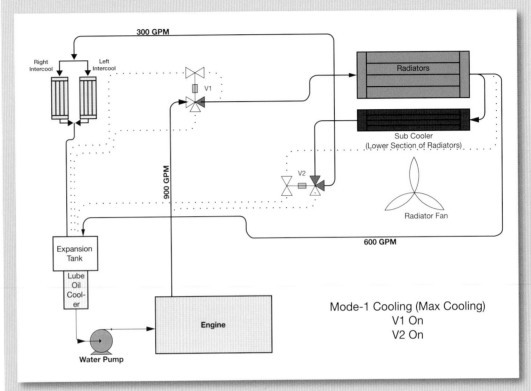

Image Credit: © Norfolk Southern Corp.

This diagram depicts a GE Dash 9 locomotive split-cooling system in maximum cooling mode. Here, 900 gallons per minute (gpm) of cooling water flows into the primary cooling coils of the locomotive's radiators. Once cooled by these coils, this volume of cooling water splits into 600 gpm, which flows to an expansion tank leading to the lube oil cooler and diesel engine, and 300 gpm, which flows to the lower "sub cooler" section of the radiators, where it is further cooled and then directed to the engine's combustion air *intercoolers*. Upon exiting the intercoolers, this portion of the cooling water also flows to the expansion tank to provide lube oil and prime mover cooling.

EMD SD70MAC Split-Cooling System

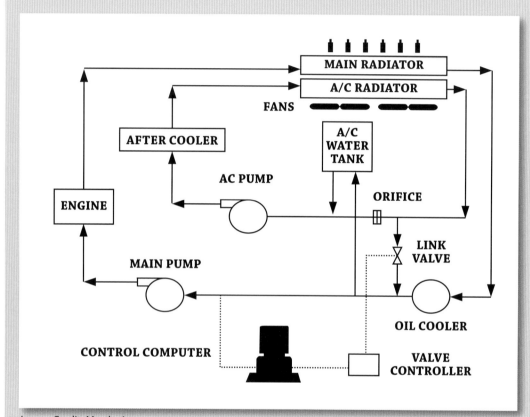

Image Credit: Martha Lenz

EMD calls this a "separate aftercooling system" because cooling water for its combustion air aftercoolers circulates in a loop which is separate from the loop that provides cooling water to the oil cooler and diesel engine. Here, each loop has its own dedicated radiator. These are stacked on top of one another with the aftercooler loop radiator receiving ambient air first. A computer controls a valve linking the two loops, so cooling water can be moved between them as needed.

Idle-Reduction Technologies

Idling has been a fact of life for the railroad industry. To cope with this reality and reduce its impact, diesel-electric locomotives typically have a low idle throttle setting. Even so, locomotives can consume 3 to 5 gallons of diesel fuel an hour while idling. The number of gallons consumed while idling climbs substantially in older locomotives or those in service in very cold climates.[55]

Prior to the introduction of products designed to eliminate unnecessary idling, the EPA estimated that road or "line-haul" diesel-electric locomotives idled an average of 38% of the time.[56] Other experts estimated that two-thirds of that idling was unnecessary.[57]

Various reasons are given for idling, including that it:

- Saves time restarting the engine
- Eliminates the risk that the engine will not start
- Maintains cooling water temperature to avoid freeze-ups
- Maintains compressed air for train air brakes
- Maintains lubrication oil temperature
- Maintains cab heating/cooling
- Maintains locomotive battery charge
- Decreases wear on diesel engine and starting systems

These are good or good-sounding reasons. Clearly bad reasons include:

- Lack of anti-idling policy
- Lack of anti-idling policy enforcement
- Engineer disinterest or lack of training

Beginning in 2000, the Environmental Protection Agency began implementing regulations that required new and remanufactured diesel-electric locomotives to reduce idling in order to reduce emissions. More stringent rules were adopted and mandated in 2008. Locomotive builders and after-market suppliers responded by developing these three technologies to address the idling problem:

- **Automatic Engine Start Stop (AESS)**, which monitors various engine systems and starts and stops the prime mover accordingly
- **Auxiliary Power Unit (APU)**, which keeps the prime mover off by providing energy for needed services from another, smaller, less energy-consuming power source
- **Wayside 480-volt Plug-in Stations,** for modified locomotives that allow battery-charging and engine coolant heating and circulation

Automatic Engine Start Stop is required by emissions rules and is the primary idle reduction technology in new locomotives today. AESS systems are simpler and less costly than auxiliary power units and don't require much space in an already crowded locomotive. AESS systems can be retrofit into older locomotives. ZTR manufactures such a system called *SmartStart IIe®*. This product saves additional energy by automatically shedding lighting and other electric loads to minimize engine restarts.[58]

An APU is a sophisticated 50 hp diesel/generator unit that typically runs intermittently or continuously once it is turned on. It monitors ambient air, locomotive cooling water and lubrication oil temperatures, brake pipe air pressure, locomotive battery charge voltage, etc., and is designed to maintain appropriate levels for these systems (e.g., engine cooling water is maintained at 100°F-120°F) while the prime mover is off. APUs have sufficient heating capacity to allow the prime mover to be shut down even when ambient temperatures plummet.

APUs are primarily used to retrofit older locomotives with idle reduction technology. For example, one APU manufacturer, HOTSTART, lists the following EMD locomotives as suitable for its product: GP8, GP9, GP10, GP11, GP35, GP38, GP38-3, GP39-2, GP40, GP40-2, SD39-2, SD40, SD40-2, SD40T-2, and SD45, as well as switcher locomotives. Other manufacturers of APU products include EcoTrans (K9™ APU) and PDI Power Drives (PowerHouse™ APU). HOTSTART states that its APU consumes one-half gallon of diesel fuel per hour and can produce annual savings of $5,000-$30,000.[59]

According to the U.S. Department of Energy's Center for Transportation Research (Argonne National Laboratory), retrofitting an older locomotive with an APU can reduce fuel use by as much as 90%, NOx emissions by 91%, unburnt hydrocarbons by 94%, carbon monoxide by 96%, particulates by 84%, and noise by 88% compared to idling the prime mover. Noise, often a community relations issue for railroad yards near neighborhoods, can be reduced by idle reduction devices or rigid enforcement of manual shutdown policies.

The EPA also claims APU and AESS systems decrease engine maintenance and extend engine life. One tradeoff regarding APUs is that its diesel engine introduces additional maintenance opportunities and costs, especially as the APU engine ages. On the plus side, compared to AESS system, an APU eliminates additional prime mover starting events, which may lengthen the operating life of engine starting motors on EMD units.

Some railroads operate in very cold climates with extended winters. Under those circumstances, idle reduction systems are less effective. The Alaska Railroad took another approach to curtail idling; it made room inside its heated locomotive repair shops for some of its

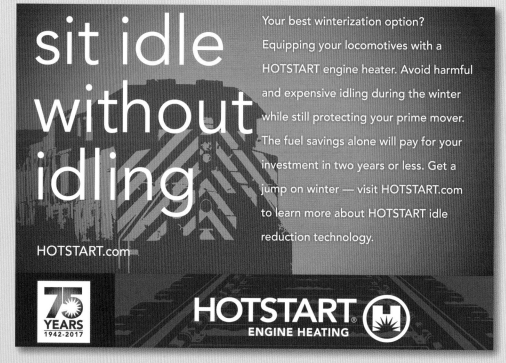

On the left, a 1993 *Railway Age* HOTSTART advertisement, and, *on the right,* one from 2017. Image Credits: HOTSTART, Inc.

HOTSTART Auxiliary Power Unit (APU).

road-ready locomotives—a clever strategy, admittedly impractical for larger railroads.

Unnecessary idling is also a problem for passenger trains that have head-end power requirements when parked at a terminal or in a yard while being serviced. Here, idling can be avoided by plugging the locomotive and its train into wayside power stations which are sometimes called *shore power*—a naval term referring to local grid power. This strategy is emissions- and noise-free at point of use.

Prime Mover Energy Efficiency Opportunities. Federally funded research conducted by Caterpillar and Cummins on more fuel-efficient, cleaner heavy-duty diesel truck engines has identified diesel engine design strategies thought to be capable of eventually achieving 50% to 55% or higher engine thermal efficiency. Moreover, as early as 2003, the U.S. Department of Energy (DOE) was projecting that diesel truck engine efficiencies could eventually reach as high as 63%.[60] While some truck engine efficiency strategies could apply to locomotive prime movers, successfully adapting these technologies to these larger medium-speed diesels will require ongoing research and development.

Another excellent source of information on potential locomotive prime mover energy efficiency improvement strategies is a 2012 United Kingdom Department of Transport study, entitled *GB Rail Powertrain Efficiency Improvements*, which was prepared by consultant companies TRL and Ricardo.[61] While the focus is on passenger and freight rail transport in Great Britain, with more attention to passenger rail, many of this study's recommendations appear to be transferable to U.S. locomotives.

Efforts to improve the energy efficiency of locomotives must take into account their harsh working environment and need for very high reliability. Norfolk Southern SD40E, Gallitzin, PA, 2013. Photo by author.

All considered, here are some prime mover energy efficiency strategies that have been, are being, or could be further explored:

- Improve fuel injection by optimizing all parameters, including injection timing, fuel pressure, fuel/air mixture, flame propagation, and burn pattern for different operating conditions—eventually achieving individual in-cylinder combustion control and multiple injection events per cycle

- Develop variable orifice diameters for intake and exhaust engine ports while reducing flow restrictions in engine intake and exhaust manifolds

- Further improve cylinder chamber geometry and piston head shape

- Develop Homogeneous-Charge Compression Ignition (HCCI) engines, which combine the best attributes of spark ignition and diesel combustion engines, to address emissions issues without the degradation of fuel economy caused by emission control technologies[62]

- Reduce heat rejection to engine block, coolant, and lubricating oil through the use of air gap pistons, exhaust port liners, and low thermal conductivity materials and coatings

- Reduce friction between engine's moving parts by improving piston ring and bearing designs and by developing lower friction lubricating oil

- Use special fuel "friction modifier" additives to reduce engine friction while producing cleaner combustion

- Improve exhaust gas utilization and energy recovery with advanced turbocharger designs, compound turbocharging, turbochargers that are mechanically connected to the driveshaft, and electric turbochargers, which are electrically driven superchargers paired with conventional turbochargers

- Explore strategies for waste heat energy recovery with organic Rankine cycles, which use water or other low global warming potential working fluid to recover exhaust or cooling water waste heat to power turbine/generators that produce additional electricity for locomotive traction motors and auxiliaries (see Section 6.2).

- Fine-tune split-cooling systems to achieve most efficient engine, combustion air, and oil temperatures while minimizing cooling system parasitic loads

- Use cooling system additives that reduce water surface tension, improving thermal conductivity, in order to increase the efficiency of heat transfer through radiators

- Use advanced materials with high heat transfer rates, like graphite foam, in primary radiators and other heat exchangers

- Explore the use of solid-state thermoelectric generators[63] on prime mover exhaust pipes and exhaust stack to directly convert exhaust waste heat energy to electricity that could be used to partially power auxiliary equipment

- Explore the use of "ram air" produced by the locomotive's movement to cool radiators instead of relying entirely on fan-driven cooling—recognizing the difficulties of applying ram air to locomotives compared to automobiles and trucks.

- Improve the efficiency of motors, pumps, compressors, fans, and

What's the Right Amount of Horsepower?

Union Pacific has historically sought very high levels of locomotive horsepower to move its transcontinental freight trains as quickly as possible across the vast expanses and changing elevations of the American West. It was no wonder, then, that as UP was retiring its 10,000 hp gas turbines in the late 1960s,[64] it ordered the highest horsepower single-unit diesel-electric locomotives ever built. These 6,600 horsepower brutes from General Motors EMD were dubbed DDA40X class locomotives. The double Ds referred to the locomotive's two four-axle trucks, and the "X" stood for experimental.

Forty-seven DDA40Xs were produced for UP between 1969 and 1971. The high horsepower of these locomotives was the product of two identical EMD 3,300 hp prime movers mounted on one 98-foot frame. But the DDA40X experiment did not last long. By 1985, all but one of these locomotives was retired in favor of newer, more reliable, fuel-efficient, lower horsepower locomotives. The exception, DDA40X #6936, has been maintained for occasional business and excursion use by Union Pacific.

In the 1990s, GE and EMD engaged in a broader industry-wide horsepower race. GE 6,000 hp AC6000CW locomotives were sold to CSX and Union Pacific, and EMD 6,000 hp SD90MAC-H locomotives were sold to Union Pacific and Canadian Pacific railroads.[65] The locomotives provided more acceleration and tractive effort at higher speeds, and therefore were best suited to faster intermodal freight trains. Intermodal trains carry large rectangular shipping containers—typically double stacked—which can be easily transferred between shipping modes, i.e., ship, train, and truck.

Higher horsepower locomotives could conceivably save energy by reducing train weight. That would occur whenever two 6,000 hp locomotives replaced three 4,000 hp locomotives—saving the weight of one locomotive (over 400,000 pounds). If high horsepower locomotives reduce the number of locomotives railroads need to own and maintain, they would also produce ownership cost savings.

The horsepower competition ended, however, when Class I railroads decided that super-high horsepower locomotives were more of a liability than an asset. They were less flexible in terms of assignment to train service, and each time one was sidelined, a full 6,000 hp of motive capacity was made unavailable for use. The DDA40X locomotives had slippage issues and were unable to fully use their high horsepower when starting or operating at lower speeds. And the AC6000CW and SD90MAC-H had engine mechanical problems.

Union Pacific's EMD DDA40X "Centennial" No. 6936 powers a westbound engineering special at Crescent, Colorado, on October 4, 2001. Photo Credit: Mike Danneman.

drives that serve the prime mover, and operate these appliances in variable flow mode where not already doing so

- Further reduce unnecessary idling

2.6 Traction and Other Alternators / Rectifiers

This section discusses locomotive traction alternators and rectifiers, which serve the diesel-electric locomotive's traction motors, as well as the locomotive's "auxiliary" and "companion" alternators that serve other locomotive functions.

An alternator is a type of electrical generator that produces AC current. Compared to a DC generator, an alternator is simpler and smaller for a given output. It also requires less maintenance and is more reliable because it doesn't have a commutator or brushes. For these reasons, diesel-electric locomotives have incorporated AC alternators instead of DC generators since the diesel locomotive's early days. An AC alternator generates power by spinning a rotor, whose "field windings" are energized or excited by direct current, within a stationary stator whose "power windings" produce the alternating current output. The three-phase AC power output of an alternator is a function of the rotor's rotational speed (rpm) and the excitation field current (amps).

Traction Alternator and Rectifiers
The locomotive's 18,000-pound traction alternator converts the mechanical energy of a rotating diesel engine crankshaft into AC electrical energy. This energy is then converted to DC electricity by a solid-state bank of rectifiers (e.g., 9 rectifiers in GE's ES44DC).

Rectifiers use semiconductor diodes to convert AC to DC. Diodes permit current to flow in only one direction. When diodes are connected to an AC power source, they produce a DC output that is choppy in appearance, representing only half of the AC sine waveform—the half when voltages are positive. In order for the rectifiers to recover all the AC power, they must also rectify the negative voltage portions of the AC waveform. A full-wave bridge rectifier circuit is used to flip these negative portions over to the positive side.

A full-wave bridge circuit designed for an input of single-phase alternating current consists of four diodes arranged in a pattern to produce an output of pulsed direct current. In contrast, a bridge circuit designed to rectify three-phase AC uses six diodes. Since the DC current produced by such a bridge is a composite form created by rectifying all

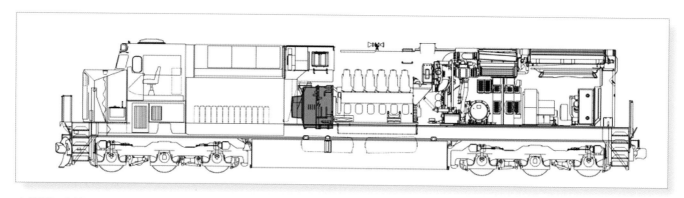

A GE ES44DC locomotive's alternator identified here as the purple device to the left of the prime mover. The alternator housing contains both traction and auxiliary alternators. Image Credit: GE Transportation.

On the left, replacement EMD alternator stator awaiting a rotor, and, *above*, a rectifier bank assembly with 2400v diodes for an EMD AR11 alternator. Both at Norfolk Southern Juniata Locomotive Shop, Altoona, PA. Photos by the author

three phases of the AC input current (which are 120 degrees out of phase), the DC output has a more or less constant DC voltage.

This DC current is conveyed from the rectifiers to the inverters by the "DC Link," which consists of positive and negative

electrical "bus bars." These bars are heavy-duty conductors—metal strips or cables—appropriately sized to handle high current.

Auxiliary and Companion Alternators

With the exception of Tier 4 locomotives, modern diesel-electric locomotives have

two or three alternators—a traction alternator and one or two secondary alternators. GE alternators consist of one unit or housing with multiple fields (electrically separate windings) serving different functions, while EMD alternators consist of physically separate alternators.

All locomotive alternators are mounted on a common shaft or gear train connected to the crankshaft of the diesel engine. While the traction alternator produces power solely for the locomotive's traction motors, the secondary alternators produce power to excite traction alternator field windings and to power auxiliary functions. These functions include fans, blowers, pumps, the air compressor, control circuits, battery charging, lighting, and cab heating and cooling (see Section 2.10).

GE and EMD use different electrical strategies. A GE ES44AC locomotive, for example, has only one secondary alternator which GE calls an *auxiliary alternator*. This alternator has three independent 3-phase outputs. The DC-rectified output of this alternator is used to charge batteries, supply power to all auxiliary functions (including lighting and control circuits), and to excite its field windings and those of the traction alternator.

In contrast, an EMD SD70MAC has two secondary alternators—which EMD calls an *auxiliary alternator* and a *companion*

A single-phase AC full-wave bridge circuit. *On left,* meter shows AC input waveform. *On right,* meter shows DC output waveform. Rectifiers on locomotives are three-phase. Image Credit: Tony Kuphaldt/Robert Hochberg.

DC voltage output

AC voltage source

Load

"Slip rings" on GE GMG197 alternator for Dash 9-44CW locomotive. These rings use brushes to make continuous electrical connections with the alternator's rotating armature. Their purpose is to deliver DC excitation current to the alternator's armature field windings. Photo by author.

alternator. These secondary alternators divide up the tasks of supplying power for auxiliary functions and alternator field windings excitation.[66] EMD's three-alternator design became common in the early days of diesel-electric locomotives (1950s and 1960s).

The manufacturers' long-standing two/three-alternator design philosophies came to an end with late model ES44AC and SD70ACe locomotives, though the change to one alternator for all electrical functions became more widely known when GE began describing it as a feature of its Tier 4 locomotive in 2015. The switch to one alternator may have been both a space-saving step and an energy conservation measure.

Magnitude of Power Production

A locomotive's traction alternator can produce a lot of electrical power, as evidenced by this comparison: If a locomotive's diesel engine supplied the traction alternator with its full 4,400 traction horsepower, the output of the alternator/rectifier would be roughly 3 megawatts. This output is enough electricity to meet the electrical needs of nearly 2,500 American households, based on a national average of 911 kWh per month. Of course, if those households were conserving and reduced their energy consumption to European levels, i.e., half that of U.S. levels, the locomotive's electrical output could serve nearly twice as many households. (Thus, this illustration of *energy production* also illustrates the power of *energy conservation*.)

Alternator/Rectifier Energy Efficiency Opportunities. The alternator and rectifiers of a modern diesel-electric locomotive have a combined design efficiency of 96% to 97%. But the fact that they require blowers to cool them indicates that they still give off a substantial amount of waste heat. Thus, there should be opportunities for energy saving through improved design and materials. As mentioned with reference to Tier 4 locomotives, eliminating the auxiliary alternator appears to have improved overall energy efficiency.

2.7 Traction Inverters

Inverters are highly efficient semiconductor devices that receive DC current as an input in order to produce a highly controllable variable frequency/variable voltage AC output. Electrical inverters are needed to drive AC motors. By varying the frequency, the speed of the motor can be varied. By varying the voltage, the motor's power output can be varied.

Inverters use large insulated gate bipolar transistors. *IGBTs* are high-powered semiconductor switches that are capable of chopping and reformulating their DC inputs into variable frequency/variable voltage alternating current. Inverters have become progressively smaller and more efficient as their technology advanced from earlier gate turn-off (GTO) thyristor semiconductors, which were used on early AC locomotives.

Power distribution circuits within the locomotive's propulsion system transmit the inverters' AC output to the locomotive's AC traction motors. Other circuits allow the motors to be reversed—so the locomotive can move in both directions. These circuits are also configured to allow dynamic braking by cutting off power to the motors while energizing their field windings. This turns motors into generators that apply resistive force to their shafts.

Until very recently, EMD and GE had different traction inverter strategies for their two-truck/six-axle/six-motor diesel-electric locomotives. EMD's locomotives had a total of two traction inverters—one per truck, or, in other words, one inverter for every three electric traction motors. In contrast, GE had and continues to use six traction inverters, one for every traction motor.

These advantages have been attributed to the GE approach:

- By providing individual control of each traction motor, slippage control is improved and traction is maximized.

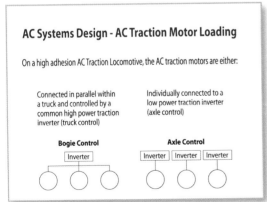

AC Systems Design - AC Traction Motor Loading

On a high adhesion AC Traction Locomotive, the AC traction motors are either:

Connected in parallel within a truck and controlled by a common high power traction inverter (truck control)

Individually connected to a low power traction inverter (axle control)

Bogie Control

Inverter

Axle Control

Inverter | Inverter | Inverter

EMD's traditional inverter/motor configuration is shown on the left while GE's is shown on the right. The term "bogie" is the British name for a railcar truck. Courtesy of former General Motors Electro-Motive Division, now part of Progress Rail, A Caterpillar Company.

- The failure of one inverter out of six idles only one traction motor, allowing five traction motors to continue operating (compared to idling three traction motors in the event of an EMD inverter failure).
- By properly sizing the inverters, if one fails, the full power of the alternator can still be directed to the five remaining traction motors—thus, maintaining tractive effort.

These advantages were attributed to the EMD approach:

- Controlling three motors in unison resists single-axle slippage by any one motor.
- Wheel-wear is equal on all truck wheels when slippage does occur.
- Manufacturing and maintenance costs are less for a locomotive with two traction inverters compared to one with six.

On its Tier 4 locomotive, EMD joined GE in having six traction inverters—one per traction motor.

Traction Inverter Energy Efficiency Opportunities. Diesel-electric locomotive traction inverters are very energy efficient mature technologies. Nonetheless, they still require cooling, which, again, is tangible evidence of energy waste. Improved design and materials could produce efficiency improvements.

2.8 AC Traction Motors

Electric motors used to propel locomotives are called traction motors because they provide pulling or "tractive" force to move an object or vehicle. A six-axle, 4,400 horsepower locomotive is equipped with six 750-800 horsepower traction motors.

AC electric traction motors are very different from DC traction motors. A DC traction motor develops mechanical power when high-voltage DC current is applied to its armature. This causes the armature to turn within fixed magnetic fields. The higher the voltage, the faster the rotation. Carbon brushes are used to conduct electrical power to the rotating commutator and armature in a DC motor.

In contrast, an AC motor develops mechanical power when AC current is applied to the 3-phase windings of its stationary stator, developing—in effect—several rotating north and south magnetic poles that induce a current and magnetic field in its rotor's conducting bars. The rotor's magnetic poles then "chase" the stator's rotating poles, causing the rotor to turn and develop torque or rotational force. In an AC induction motor, the stator's and rotor's magnetic poles are never perfectly aligned; there is always a percentage of slip between them. This slip is greater when the motor is under load, drawing more current and operating more efficiently.

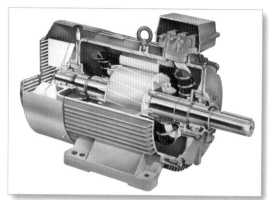

Cutaway diagram of an AC motor. Image Credit: Gibbons Engineering Group.

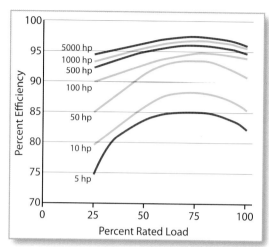

Motor horsepower curves. High hp motors achieve 95+% efficiency. Image Credit: U.S. Department of Energy

Because its rotating element operates by induction, an AC motor does not have a commutator or brushes.

Diesel-electric locomotives used DC traction motors for many years in large part because controlling the speed of AC motors for traction purposes was so difficult.[67] Voltage and frequency must be controlled precisely for any given rotational speed. If either is off slightly, the motor may burn up. It was not until the 1990s that functional inverters (described in the preceding section) combined with microprocessor controls were available to effectively control large AC motors in variable speed traction applications.

The adoption of AC traction motors was initially slow because locomotives equipped with them cost more. However, virtually all line-haul diesel-electric locomotives purchased today are equipped with AC electric traction motors.

Superiority of AC Traction Motors

There are good reasons why new diesel-electric locomotives are being equipped with AC motors. Their features include:

- Improved durability with less maintenance because they are simpler in design, with fewer moving or rubbing parts (e.g., they have no brushes), and are not subject to commutator "flashovers"[68] that can destroy armatures, windings, and other parts
- More precise control and smoother application of torque—increasing adhesion and ultimately horsepower delivery to the rails
- Capacity for very slow speed operation under load (and even holding a train on a hill at standstill) without overheating—unlike DC traction

motors, which must operate at minimum rotational speeds to avoid overheating
- Reduction in the number of locomotives required to pull heavy trains or serve as helpers in mountainous terrain
- Improved dynamic braking capacity and range

An interesting historical note: It took the advent of AC traction to produce a single diesel-electric locomotive finally capable of exceeding the starting tractive effort of the aforementioned Norfolk & Western Y6 locomotive.

While admittedly not representative of many other steam locomotives, the Y6b and Y6c variants of this workhorse were traction champions capable of delivering 152,000 and 166,000 pounds of starting tractive effort to the rails, respectively.[69] Demonstrating the power of AC traction motors 25 years ago, EMD's SD60MAC achieved a starting tractive effort of 175,000 pounds.

Electric Traction Motor Energy Efficiency Opportunities. Traction motors are regarded as a mature and efficient technology. However, even 96% efficient AC motors run hot and require substantial active cooling to remove waste heat. Again, this suggests opportunities for further efficiency improvement through improved design and materials.

Additional traction motor efficiency might be achieved by reducing the conductivity of motor stators and rotors (to reduce resistive and rotor slip-related losses), upgrading the quality of steel used in the motor (to reduce hysteresis losses), and improving the insulation between thinner winding laminations (to reduce eddy current losses).[70]

EMD SD60MAC demonstrator street-running in Bellevue, Iowa, October 17, 1993. Photo Credit: Mike Danneman.

2.9 Locomotive Trucks and Running Gear

The diesel-electric locomotive's two trucks contain wheels, axles, traction motors, gears, springs, dampeners, bearings, brakes, and other running gear. The trucks must:

- Support the weight of the locomotive and its electric traction motors
- Enable even application of power to the rails
- Transfer the forces of the traction motor wheelsets to the locomotive platform

N&W Y6 in Iaeger, WV, October 24, 1959. Note massive low-pressure pistons. Photo Credit: Benjamin T. Young, Collection of Outer Station Project.

- Provide smooth and stable handling for all track conditions
- Improve fuel economy by minimizing rolling resistance and improving rail adhesion when needed
- Minimize damage to rails and track
- Provide a comfortable ride

The only diesel-electric line-haul freight locomotives that GE and EMD build for use at this time in the U.S. market sit on two three-axle, six-wheel trucks. The nomenclature for this truck/axle arrangement is C-C.

Many older diesel-electric locomotives were B-B, also having two trucks but each with two axles and four wheels. Many of these locomotives are still in use, often rebuilt and repowered by railroads and put to work as yard switchers or slugs, which are discussed in Section 2.14. Also, modern diesel-electric passenger locomotives are all B-B.

In supporting the locomotive's massive weight, which in the case of line-haul freight locomotives exceeds 400,000 pounds, trucks distribute the weight so that individual axle loadings do not exceed approximately 70,000 pounds. This is the weight that general service main-line track—and especially bridges—can support without damage.

Attached to the bottom of the locomotive frame are two large structures called *traction pins*. Each pin mates with a circular

GE Hi-Ad Truck on Norfolk Southern C44-9W locomotive. Coil springs and roller bearings at axle ends are evident. Photo by author.

GE Evolution locomotive 3-axle truck assembly with 3 nose-suspended AC traction motors. Photo Credit: Sean Graham-White.

Timken low-friction HDL seal bearing on replacement axle/wheelset. Photo by author.

the top of a truck that accepts and applies the weight of a locomotive to the rest of the truck.

The traction pins permit the trucks to pivot, so that the locomotive can follow curves, and are strong enough to handle the forward and rearward forces the truck delivers to the locomotive frame when all traction motors are operating at full output. But the pins do not support the weight of the locomotive. Weight is supported by rubber/steel plate compression pads, which GE calls *load bearers*. Earlier locomotives used a different system for mounting locomotive frames onto the trucks. A round *center plate* on the frame

opening in the truck's bolster. The bolster is the transverse steel structure on

would fit within and mate with a round deep-dish-like *center bowl* on the truck bolster—with the weight of the locomotive "securing" the frame to the trucks.

Trucks also house tapered roller bearings mounted on the ends of the axles that allow wheelsets to turn with minimum friction. Needless to say, given the forces that locomotive wheels encounter, their hardness and strength are of great importance and have been subject to metallurgical improvement over the years.

Locomotive trucks are heavy. Each one weighs nearly 40,000 pounds, with almost half that sum made up by the

Wheel/axle assemblies with large bull gears that engage smaller motor pinion gears. Photos by author.

Wheel/axle/motor assemblies awaiting installation on rebuilt locomotives. Norfolk Southern Juniata Locomotive Shop, Altoona, PA.

weight of the electric traction motors. These motors are aligned with their armatures parallel to the driven axles and *nose suspended*, i.e., attached to the driven axles by two bearings and to the truck at a single point by a bracket called the *nose*. This arrangement allows the smaller pinion gear of the motor to rest on and engage a larger bull ring gear attached directly to the axle. It also leaves a portion of the motors' weight *unsprung*, i.e., not supported by the trucks' springs. Unsprung weight is a detriment to the handling of the truck. It increases the locomotive's impact on the track, especially in curves, on poorly aligned track, at higher speeds, and when trucks "hunt" between the rails.

Hunting is the oscillation of wheel flanges against inside rail surfaces. While some hunting is unavoidable, given rail tolerances and the tapered profile of locomotive and railcar wheels, excessive hunting is destructive. It is also energy wasteful because wheels oscillating back and forth between rails increases rolling friction. This slows down a train or requires more locomotive horsepower to pull it at any given speed.

Top speed is a function of the locomotive's gear ratio—the ratio of the number of teeth on larger bull ring gear to the number of teeth on the smaller pinion gear. Typical freight locomotive gear ratios producing 70 to 75 mph top speeds are 70:17 or 83:20. A higher gear ratio produces greater mechanical advantage and, all other things being equal, a higher tractive effort with lower top speed. The gear assembly is enclosed in a sealed metal case that contains special lubricating oil to reduce friction.

Springs and dampeners in the trucks cushion the locomotive over track irregularities and minimize side-to-side oscillation. This suspension smoothes and stabilizes the operation of the locomotive on the rails—thus, enhancing safe operation; improving operator comfort; protecting locomotive equipment; decreasing forces to the rail, track damage, and wheel wear; and saving energy by reducing wheel/rail friction. Locomotive trucks are also equipped with compressed air-actuated friction brakes that press brake shoes against wheel treads to slow or stop.

While the vast majority of new diesel-electric locomotives are built with traction motors powering all six axles, GE and EMD now both sell special six-axle locomotives with only four powered axles (two on each truck). These locomotives are designated the ES44C4/ET44C4 and SD70ACe-P4/SD70ACeP4-T4, respectively (the latter GE and EMD models being Tier 4 compliant).[71] While the middle axle on each truck is unpowered on the GE locomotive (an A1A-A1A configuration), the unpowered or idler axle is inboard closest to the fuel tank on the EMD (B1-1B).[72]

These designs save money by eliminating the cost of two AC traction motors and their accompanying electronics—in effect, producing an AC-powered locomotive for nearly the same price as a DC locomotive. These locomotives cannot produce the tractive effort or pulling ability of locomotives with six AC motors (and thus are not as well suited for hauling heavy trains on mountainous routes), but they can match the pulling ability of late model locomotives with six DC motors. They also possess the other advantages associated with AC motors. Because the C4/P4 locomotives have only four traction motors per locomotive, 25% of their power is lost when one motor fails (instead of 16.67%).

GE and EMD C4/P4 locomotives are designed to transfer weight from their unpowered axles to their powered ones when additional adhesion is needed.[73] This is accomplished by raising the unpowered axles. However, this action can increase axle loading to more than 100,000 pounds, which is far in excess of the normally permissible 70,000 pounds.

Trucks and Running Gear Energy Efficiency Opportunities. The primary sources of energy consumption in locomotive trucks are their electric traction motors, whose efficiency opportunities were discussed earlier (see Section 2.8).

In the operation of locomotive trucks we see that friction is friend and foe. Conceptually, additional energy could be saved by developing new or improved strategies and technologies that increase friction when it is needed for adhesion purposes and decrease friction when it is not helpful, and is, in fact, just holding back the locomotive and its train.

Further advances in roller bearing systems—with improved geometry and seals—might be able to further reduce axle friction in locomotives as well as in the railcars they pull. Previous improvements in bearing seal design by the Timken Company and Burlington Northern Railway in the 1980s are said to have reduced "trailing tons"—the pounds of resistive force associated with pulling a train on level track—by as much as 5% at 50 mph and 14% at 10 mph.[74] More recent improvements have produced an additional 2% to 4% friction reduction.[75]

The Challenge of Adhesion

The low coefficient of friction between rolling steel wheels and steel rails, and the small contact patches between those wheels and rails, have pros and cons. They provide railroads with a significant energy efficiency advantage because locomotives and railcars roll so easily but are a detriment to locomotives when they need maximum traction to get a train started or when ascending grades. It's lack of

adhesion, not horsepower, that limits the amount of force a locomotive can apply to the rails at standstill and slower speeds.

Needless to say, adhesion and avoiding wheel slippage have been issues for the railroads ever since locomotives were powerful enough to deliver more force to the rails than frictional forces could contain. Steam locomotives were especially prone to wheel slippage because their pistons, drive rods, and drive wheels delivered uneven, pounding forces to the rails. Putting more weight over the driving wheels helped, as did a skilled engineer.

The amount of tractive effort a diesel-electric locomotive with DC motors can apply to the rails under optimal track conditions is approximately 25% to 30% of the weight it places on its drive wheels. This fraction is referred to as the *adhesion factor*.

The contact area a single locomotive wheel has with the rail is less than the size of a dime. It is through the miniscule area of less than 12 dimes—much smaller than a human hand—that modern diesel-electric locomotives exert as much as 200,000 pounds of tractive force. Photo Credit: James Ulrich.

Locomotives with AC traction motors have higher adhesion factors. When EMD demonstrated its SD60MAC locomotive in the early nineties, it reported a starting adhesion factor of 45%, and a "dispatchable" adhesion factor of more than 35%.[76] This was a remarkable achievement—adhesion factors comparable to today's AC-motored locomotives.

The most modern diesel-electric locomotives produce as much as 200,000 pounds of starting tractive effort and 175,000 pounds of continuous tractive effort. Successfully delivering that force to the rail requires adhesion. The technologies that permit locomotives remarkable adhesion include:

- Sensors that identify wheel slippage by comparing wheel rotational speed to true ground speed (using radar) or by comparing the current draw of all traction motors to identify a motor that suddenly shows a drop in current and a spike in voltage
- Computer systems that control power to motors to prevent or stop wheels from slipping
- Inverter-controlled AC traction motors that are capable of very slow, well-controlled operation
- Hi-Ad (high-adhesion) trucks that redistribute the weight of the locomotive on the trucks to counteract the unloading of the front axles on each truck when the locomotive is under load[77]
- "Wheel creep" systems that are designed to maximize friction and adhesion by allowing the locomotive's wheels to slip slightly (i.e., up to a few mph) in a highly controlled fashion when applying power or when using dynamic brakes
- Self-steering *radial* (EMD) or *steerable* (GE) trucks that follow track curvature by permitting the first and third axles of three-axle trucks to slightly pivot while the middle axle moves laterally
- Automatic rail sanding and compressed air systems, where the latter blow moisture and debris from slippery rails
- Adding weight to locomotives (i.e., ballasting)

Diesel-electric locomotives, as early as the 1940s, had wheel slip detectors that worked by comparing DC electric traction motor voltages.[78] If one motor had a much higher voltage than the other (on a two-axle truck), a wheel slip indicator light and/or buzzer would tell the engineer to ease up on the throttle. Since then, diesel-electric locomotives have adopted increasingly more sophisticated technologies to avoid wheel slip – initially described as "wheel slip control," and now "adhesion control." The EMD SD40-2 locomotive manufactured in 1980, for example, was equipped with a wheel slip control system that quickly and automatically would reduce power and apply sand to the rail when wheel slippage was detected.[79]

Wheel creep control increases adhesion by allowing a small amount of wheel slippage even though incrementally higher levels of slipping would produce a loss of traction. With AC motors, the exact percent of "creep" can be dialed in. Wheel creep may cause an audible chirping sound as traction is lost and regained. This causes additional wear on wheels, and to a lesser extent on the rails, but has been partially offset by improved metallurgy. Wheel creep technology was first introduced by GM's EMD over 30 years ago.[80]

Radial or self-steering trucks align wheel flanges with the rails—reducing lateral forces and flange wear, improving fuel economy, and substantially increasing adhesion in curves. Radial-type trucks became important for three-axle trucks, given their longer fixed wheelbases and other features. The EMD radial truck, whose current model is the HTCR, was more successful than GE's. The success of EMD's radial trucks is evident in the fact that all EMD locomotives purchased by Norfolk Southern after 1993 have been equipped with these trucks. However, timely overhaul of these trucks is essential to avoid their imposition of abnormally high lateral forces on the rail.

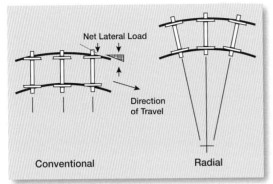

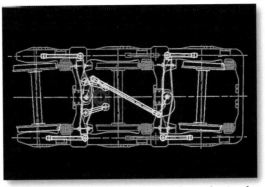

On left, "'reduced net lateral load" on wheel flanges when axles pivot to align wheels with curve. *On right,* mechanism for pivoting axles in EMD radial truck. Courtesy of former General Motors Electro-Motive Division, now part of Progress Rail, A Caterpillar Company

EMD's compressed air system is called *Railblaster*, and GE's is called *Advanced Rail Cleaner* or *ARC*.[81] These truck-mounted devices blow high-pressure air on the rails just ahead of the wheels on the locomotive's lead axle. Clean, dry rails enable heavier trains to run with fewer locomotives. However, this use of compressed air has an energy penalty because it increases air compressor operation.

All other things being equal, heavier locomotives can successfully apply more tractive effort to the rails than lighter ones. For this reason, railroads concerned about low-speed tractive effort have elected to operate heavier locomotives. The 70,000-pound axle-loading restriction has limited diesel-electric road locomotives to 420,000 pounds, though there have been exceptions, e.g., the GE locomotives manufactured for Union Pacific (427,000-428,000 pounds) and CSX and Norfolk Southern (432,000 pounds). While these heavier locomotives are capable of achieving

higher levels of adhesion, they require more fuel to propel their additional weight and may need upgraded truck design or components.[82]

Improved adhesion technologies save energy not only by avoiding the direct fuel waste associated with instances of runaway slippage but also by improving railroad "asset management." Improved adhesion allows railroads to move trains more efficiently over their networks, minimizing stop-and-go operation. And since fewer locomotives are required to pull those trains, each train has fewer locomotives idling when idling is unavoidable.

CSX ES44AC-H leading long intermodal train on former New York Central mainline through Buffalo, NY, November 2013. Photo by author.

Use of Rail Lubricants and Friction Modifiers

Locomotives pulling trains up gradients benefit from top of rail (TOR) *friction modifiers* that fine-tune wheel/rail friction to a specific coefficient to improve locomotives traction.[83] On the other hand, unwanted wheel/rail friction can be minimized through the use of gauge face (GF) lubricants; these biodegradable, nontoxic products make the inside surface of the rail—where wheel flanges make contact—more slippery. This reduces the resistive force locomotives must contend with, especially on curving tracks.

Rail friction modifiers and lubricants are usually applied by wayside applicators that may utilize solar panels as an energy source. TOR friction modifiers can also be deployed by applicators mounted to locomotive trucks. In addition to reducing fuel consumption, friction modifiers and lubricants reduce wheel and rail wear, rolling contact fatigue, and wheel climb derailments. They also increase the time interval between rail grindings that ensure an optimal rail profile.[84] Fuel savings are estimated at an average of 2% to 7% (with as much as 10% on curving mountainous routes operated at 45 mph).[85] Energy savings derive from reducing train rolling resistance and improving the ability of the locomotive to deliver power to the rails.

Foam bar applicator for water-based top of rail (TOR) friction modifiers. Photo Credit: L.B. Foster.

Wayside applicator for gauge face (GF) rail lubricants. Photo by author.

CSX ES44AC-H exits tunnel and crosses the Upper Potomac River on approach to Harpers Ferry, WV, June 2014. Photo by author.

2.10 Auxiliary Energy Systems

The auxiliaries are energy-consuming equipment other than the traction alternator, rectifiers, inverters, and motors. As such, they include:

- Cooling fans
  - Diesel engine radiator fan(s)
  - Traction motor blowers
  - Alternator blowers
- Pumps for diesel fuel, lubrication oil, and cooling water
- Air compressor providing 130 psi compressed air to two main air reservoirs or tanks (one serving locomotive functions and the other the train brake—after appropriate pressure reduction)
- Electronic equipment power supplies
- Battery charging
- Lighting
- Cab heating, ventilating, and air conditioning (HVAC)
- Air filter exhaust fan(s)
- Alternator field excitation circuits

While auxiliary functions are essential to the proper operation of the locomotive, and many of them contribute to the overall energy efficiency of the locomotive, they are defined as "parasitic loads" to the locomotive's diesel engine. In other words, these loads consume some of the power of the prime mover (typically 100-300 hp), leaving less available for traction purposes. As such, they represent energy and fuel consumption penalties.

One parasitic load often not considered is the locomotive's horn. Every time it sounds, compressed air is utilized. Thus, safety rules that require more frequent use of the horn when approaching crossings actually have small energy penalties associated with them. Newer diesel-electric locomotives use electronic bells. Perhaps an energy efficient electronic horn will be developed. This is admittedly a minor point, but it demonstrates how energy use and, therefore, conservation opportunities abound in the greater and lesser features of the diesel-electric locomotive.

Top rear view of EMD F125 passenger locomotive ready for use by Southern California Regional Rail Authority's Metrolink. *On the left* is the dynamic braking grid fan, and *on the right* are the primary radiator fans. The dynamic brake fan is not considered an auxiliary or parasitic load because it is self-powered by electricity generated by the dynamic braking process itself, and not the prime mover. Photo Credit: Progress Rail, A Caterpillar Company

Auxiliary Energy Systems Efficiency Opportunities.

Reducing auxiliary loads is important because it allows a greater percentage of the diesel engine's output to be available for traction purposes. There are three fundamental strategies for reducing the energy consumption and energy penalties of locomotive auxiliaries:

- Improve the energy efficiency of the alternator(s) that supply the auxiliaries
- Eliminate auxiliary/companion alternators and instead power the auxiliaries with the larger (and presumably more efficient) traction alternator, as in late model Tier 3 and current Tier 4 line-haul freight locomotives

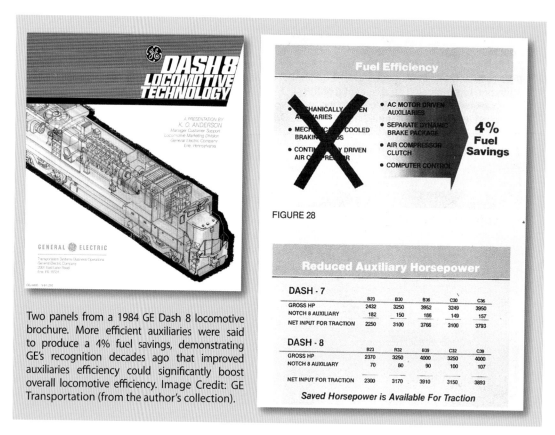

Two panels from a 1984 GE Dash 8 locomotive brochure. More efficient auxiliaries were said to produce a 4% fuel savings, demonstrating GE's recognition decades ago that improved auxiliaries efficiency could significantly boost overall locomotive efficiency. Image Credit: GE Transportation (from the author's collection).

- Reduce the auxiliary loads

The alternator used to provide power for auxiliary loads is a mature technology. But since it is fan-cooled, we know it gives off waste heat and therefore could be made more energy efficient.

There are numerous ways to reduce the amount of energy that auxiliaries consume. The efficiency of fans, pumps, and compressors can be increased by microprocessor/in-verter controls where this has not already been done. Additionally, all these devices could benefit from mechanical design improvements. Cab heating and air conditioning systems could be more efficient. Engine cabs could be better insulated. LED lighting should be used.

Air compressor energy consumption could be reduced by minimizing brake pipe and brake cylinder air leaks, and by supplanting the use of train brakes with increased use of dynamic brakes—consistent with safe train handling.

2.11 Locomotive and Train Braking Systems

Being able to slow and stop a train, and secure it once stopped, is at least as important as starting and running one. Accordingly, engineers have at their disposal four braking systems:

- Locomotive independent brakes
- Locomotive dynamic brakes
- Train air brakes (*train brake* or *automatic brake*)
- Locomotive and train parking brakes (*hand brakes*)

Locomotive and train braking systems are energy systems. They ultimately stop trains by transforming the mechanical energy of train-mass-in-motion into thermal energy. This heat is dissipated into the atmosphere.

Mechanical Energy → Thermal Energy

However, as we shall see, the actual mechanisms for accomplishing braking are complicated and involve additional energy transformations.

Locomotive Independent Brakes

Independent brakes are locomotive friction brakes. They use compressed air produced by the locomotive's air compressor to force friction brake shoes against the wheel treads of locomotive wheelsets. As previ-

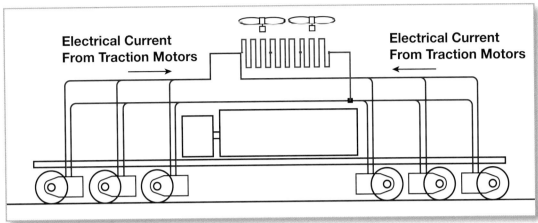

Schematic diagram showing dynamic braking grids (depicted in red) that dissipate as heat the electricity produced by traction motors when they operate as generators. Image Credit: EngineeringExpert.net, LLC.

ously explained, the "wheel tread" is that portion of the wheel that makes contact with the rail head. It's slightly tapered such that the outside of the tread is of smaller diameter and circumference than the inside of the tread closest to the flange. This tapered feature allows locomotive and railcar wheels to negotiate curves more easily—with less squealing, friction, and wear. Railroad operating rules actively discourage the use of locomotive brakes for train braking.

Locomotive Dynamic Brakes

These brakes are the primary braking system available to the locomotive engineer to slow or control the speed of a train. The advantages of dynamic braking are so substantial that railroad rules may stipulate that these brakes are the first order of braking a train under normal (non-emergency) conditions.

However, under certain circumstances, engineers may prefer reliance on train air brakes to achieve proper train handling, e.g., when negotiating undulating (*hogback*) track alignments.

Dynamic brakes work by using the locomotive's electric traction motors as generators. In dynamic braking (generating) mode, these motors exert a retarding force on the locomotive's axles and wheels while generating electricity that is dissipated as heat by the locomotive's dynamic braking grids.

While all locomotive and train brakes on diesel-electric locomotives turn mechanical energy into heat energy, dynamic brakes function by first turning mechanical energy into electricity.

Mechanical Energy → Electrical Energy → Thermal Energy

Dynamic braking grids incorporate large resistors, resembling toaster elements, that heat up when electrical current passes through them. Heat rejection to the atmosphere is facilitated by fans that are powered by electricity generated by dynamic braking. Braking grid fan operation is essential to maintaining the grid elements. Without cooling air, the grid would burn up on first use.

On GE locomotives, dynamic braking grids are typically located at the top of the locomotive a short distance behind the operator's cab. On EMD SD70M-2 and SD70ACe locomotives, these grids are located at the rear of the locomotive. This location may

have been chosen for packaging reasons, but it also had the effect of reducing crew exposure to the noise and vibration caused by dynamic braking grid fans.

While perhaps counter-intuitive, dynamic braking consumes diesel fuel. In order for the armature of a traction motor to create a retarding force, it requires field current—and this current must be increased as higher levels of dynamic braking are selected. That current is supplied by the locomotive's main alternator, which is powered by the prime mover. A typical SD70ACe locomotive might burn 4 gallons of diesel fuel per hour in dynamic braking Notch DB2 and 14 gallons per hour in DB8. However, these rates of fuel consumption are not significant. For comparison, the SD70ACe idles at 3 gal-

lons per hour and runs in Notch 8 at 187 gallons per hour.

Operating Dynamic Brakes. On locomotives with vertical control stands (which are most common in freight locomotives), dynamic braking is controlled using a separate dynamic braking lever that is employed after the throttle is placed in idle (see photo illustration in Section 2.2). When locomotives have desktop controls, where a single "combined power" lever controls both throttle and dynamic braking, the lever is first placed in idle before it is moved to dynamic braking.

Since dynamic braking only applies braking force from the locomotive, the engineer anticipates the run-in of slack and

Examples of radial and box dynamic braking resistor grids for EMD locomotives. Image Credit: Mosebach Manufacturing Company.

the resulting bunching of cars toward the locomotive as the brakes are applied. Once this has occurred, good handling practices call for the engineer to select the desired level of dynamic braking by slowly moving through DB1 through DB8 settings, or, in some locomotives, DB1 through DB9— where the higher settings apply more braking force. Once in dynamic braking,

the braking lever (or combined power lever in the braking portion of its range) moves smoothly from DB1 through DB8 or 9, permitting modulation of braking force outside of discrete notches.

Dynamic brakes are powerful, and modern locomotives can exert a level of braking force equal to nearly 80% of the propulsive force

diesel-electric locomotives can produce. With dynamic brakes engaged, it may be possible for the engineer to avoid or substantially reduce the use of train air brakes (which are explained next) for long periods of time. This can enhance train control and safety, and it produces significant economies by minimizing wear on train car brake shoes and wheel treads. Dynamic brakes also prevent car wheel damage caused by the overheating that occurs when train brakes are used for extended periods. By transferring all the braking action to the locomotive wheel/rail interface, dynamic brakes do slightly increase wear on locomotive wheels.

In a few mountainous areas, it is still possible to see trains with "helpers," i.e., additional locomotives that get temporarily attached to the rear and sometimes the front of heavy trains. These extra locomotives not only provide extra horsepower for pushing and pulling trains uphill, but they may also provide extra braking— through the use of their dynamic brakes— to slow trains as they move downhill.

More on How Dynamic Brakes Work.
Irrespective of whether a locomotive has AC or DC motors, the amount of braking force developed by dynamic brakes is a function of train momentum and the amount of electricity traction motors generate and dissipate through braking grids when dynamic brakes are applied.

Dynamic braking grid vents visible behind cab in Norfolk Southern GE D9-40CW locomotive traveling through the Allegheny mountains in SW Pennsylvania, June 2015. Photo by the author.

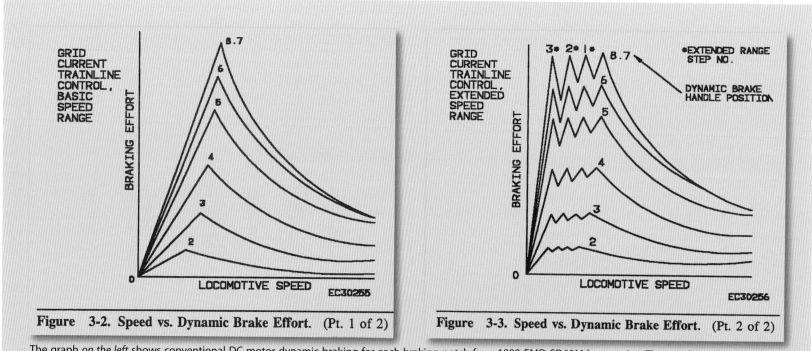

Figure 3-2. Speed vs. Dynamic Brake Effort. (Pt. 1 of 2)

Figure 3-3. Speed vs. Dynamic Brake Effort. (Pt. 2 of 2)

The graph *on the left* shows conventional DC motor dynamic braking for each braking notch for a 1992 EMD SD60M locomotive. The graph *on the right* shows the application of extended range DC dynamic braking. The jagged-tooth pattern is caused by varying braking grid resistance levels. Courtesy of former General Motors Electro-Motive Division, now part of Progress Rail, A Caterpillar Company

In a locomotive with DC traction motors, dynamic braking is initiated by disconnecting power to the motor armature while supplying current to energize or excite the field windings surrounding the armature. Then, as the momentum of the train turns the armature, it generates electricity within the electric field created by the energized field windings and delivers that electricity to the dynamic braking grids. However, as the train slows and loses momentum, dynamic braking capacity diminishes. This loss of braking can be mitigated by in-

creasing current to the field windings. But this strategy only works to a point when the maximum current handling limits of these windings have been reached. At that point—around 20 mph—DC locomotives without extended range dynamic braking lose their dynamic brakes quickly.

To address this problem, locomotive manufacturers equipped DC-motored locomotives with "extended range" and "higher capacity" dynamic braking systems. These systems increase dynamic braking capacity

at lower speeds by decreasing the electrical resistance of the dynamic braking grid in a step-by-step manner. As the resistance decreases, the electric load increases. This causes more current to flow from the motor armature, which, in turn, increases the motor's braking effect. This works well until about 9 mph, and then starts to drop off significantly. At about 3 to 5 mph, DC dynamic braking effect is reduced to zero.

AC traction motors work differently. As previously explained, they are controlled by inverters that modulate frequency and voltage—thereby controlling the slip angle between the rotating magnetic poles of the stator's field windings and the rotor. If the stator's rotating magnetic poles lead the rotor's magnetic poles, then the motor is powered and generating torque. On the other hand, if the rotor's magnetic poles lead those of the stator, the motor will re-

sist rotation and generate electricity. With inverter control, AC traction motors can provide substantial retarding force to nearly zero mph—though skillful engineers may also be able to use extended range DC dynamic brakes to slow a train sufficiently that train resistance will bring it to a stop.

Dynamic Braking Energy Efficiency Opportunities. Dynamic braking serves an important purpose and represents a major advantage diesel-electric locomotives had over steam locomotives. Sadly, however, diesel-electric freight locomotives have not been designed to make use of the power that their traction motors generate when they are in dynamic braking mode; as previously explained, this energy is just lost to the environment as heat. Thus, the principal energy efficient opportunity associated with locomotive dynamic braking is the recovery and reuse of as much of this energy as possible. One need not think further than the hybrid car that may be parked in your driveway to realize that there must be effective ways to accomplish this (see Section 6.3 for discussion of hybrid locomotives).

Some diesel-electric passenger locomotives have made or are

making use of dynamic braking electricity for HEP. The EMD F69 prototype, built in 1969, had this feature, as do the newest passenger locomotives, EMD's F125 and Siemens SC-44 Charger.

Train Air Brakes

Train air brakes are properly referred to in the singular as the *train brake* or *automatic brake*. This braking system uses compressed air supplied by the locomotive to pressurize a 1¼-inch air line that runs car to car from the locomotive to the end of the train. Technically, this line is called the *brake pipe* but is commonly called the *trainline*. By varying the pressure in this line, the engineer can apply or release friction brakes on all cars in a train, including the locomotive.

The train brake consumes diesel fuel indirectly as a result of its use of compressed air. This air is supplied by the locomotive's air compressor which runs on electricity produced by the alternator and prime mover. Thus, the train brake's energy transformations look like this, beginning with the chemical energy in the diesel fuel consumed by the prime mover:

*Chemical Energy → Thermal Energy
→Mechanical Energy →
Electrical Energy → Mechanical Energy (air compressor) →Potential Energy (compressed air reservoir) →
Mechanical Energy →
Thermal Energy*

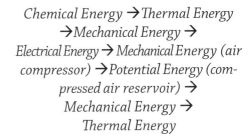

Vintage EMD print ad explaining that the dynamic brakes on the diesel-electric locomotives powering the Santa Fe "El Capitan" passenger train saved 1,700 pounds of iron brake shoes per round trip and 600,000 pounds per year. That's a lot of iron! *Saturday Evening Post*, October 27, 1951. Courtesy of former General Motors Electro-Motive Division, now part of Progress Rail, A Caterpillar Company.

On each car, the train air brake system consists of the following components:

- Control valve
- Air reservoir
- Brake cylinder
- Friction brake shoes (pads)

The system is designed so that the trainline supplies pressurized air to the air reservoir on each car. This has auxiliary and emergency compartments. When trainline air pressure is reduced, as a result of either the engineer making a train brake application or an accidental loss of air pressure, the control valve responds by directing pressurized air from the air reservoirs of each car to the brake cylinders, which, in turn, apply the friction brakes (brake shoes) against car wheel treads.

Railcar brakes are completely off when trainline pressure is approximately 80 psi for freight trains and 90 psi for passenger trains. A 6-7 psi reduction produces a minimum application of the train brake. A rapid reduction of trainline air pressure initiates emergency braking, where deacceleration is limited to the level of braking effort that just avoids wheel sliding (because sliding not only damages wheels—causing flat spots and, potentially, flange damage—but also increases stopping distance).

This system uses control valves that operate on the same principle as the triple

Locomotive and Train Air Brakes a Century Ago

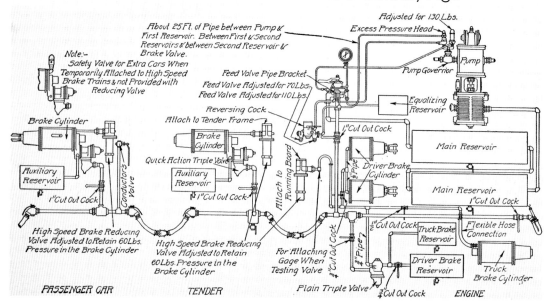

Diagram from *Air Brake: An Up-to-date Treatise on the Westinghouse Air Brake as Designed for Passenger and Freight Service and for Electric Cars, with Rules for Care and Operation*, written by Llewellyn V. Ludy in 1918. The portion of the train air brake system located in the locomotive is depicted *on the right*, with the passenger car brake system *on the left*. Since this is for a train pulled by a steam locomotive, the brake system for the tender is provided in the middle. Image Credit: American Technical Society (Public Domain).

valve design invented by George Westinghouse in 1868, 150 years ago. However, current designs are far more sophisticated, efficient, and effective.

When this braking concept was designed, it was intended to provide fail-safe braking wherein loss of air pressure for any reason (e.g., train separation) caused the brakes to be applied. Realistically, they are not completely fail safe because in order to work properly, car air brake auxiliary reservoirs must be sufficiently charged with compressed air from the brake pipe to be able

to press the brake shoes against the wheel treads. If multiple braking applications and releases are made over a short period of time, or train brakes are applied for long periods of time, these reservoirs can be discharged to a point where they lack sufficient air pressure to provide any more braking. Thus, engineers must be careful to safely conserve (and not squander) the car air reservoir supply when descending long grades if they are controlling trains with the train brake. The use of dynamic braking mitigates this risk.

One reason why train brake applications consume compressed air is that brake cylinders, when actuated, are inclined to leak air. They are more prone to leakage over time because they operate in a harsh railroad environment of vibration, impacts, and low temperature extremes. At least one brake valve manufacturer, New York Air Brake, now markets a Brake Cylinder Maintaining (BCM) brake valve that compensates for this air leakage by using trainline air to restore a claimed 85% of the original brake cylinder "target pressure."[86]

Instead of cabooses, modern trains have end-of-train devices (often referred to as EOTs or ETDs) installed on the rear of the last car. These tell engineers the air pressure in the brake pipe at the very end of the train.[87] This information helps them know whether car brakes at the end of the train are applied or released and is an indicator of whether the train will have full braking capacity when needed. EOTs communicate with the cab via radio transmissions and have motion detectors that tell the engineer whether the rear of the train is moving forward, backward, or is stopped. These devices are typically powered by batteries. However, they may also be powered by small turbine-generators that operate off of compressed air from the brake pipe—saving batteries but imposing a parasitic load.[88]

When the engineer applies the train brakes by reducing brake pipe pressure, the pressure in the line drops, but not all at once. It drops incrementally, car by car, beginning first with the cars closest to the locomotive. Thus, braking does not occur in all the cars of the train simultaneously. The locomotive engineer must anticipate the resulting bunching of cars toward the locomotive (loss of slack). Bunching describes "buff" forces (in compression), while "draft" forces describe tension in the couplers that exists when slack has been pulled out.

By releasing air and lowering train brake pipe air pressure from the locomotive as well as the EOT device, the engineer can use the EOT to assist with emergency braking. This speeds up braking response time while also reducing the forces when slack runs (diminishes) between cars and causes bunching. While dynamic brakes can reduce the use of train brakes and substantially slow trains down, train brakes are commonly used to bring a train to a stop.

A few railroads have experimented with electronically controlled pneumatic (ECP) train brake systems. These systems use a power and data line, which runs (along with the brake pipe) from the locomotive through the length of the train, to give the engineer the ability to electrically actuate all train air brake valves simultaneously. With ECP brakes, the train brake pipe still keeps brake air reservoirs pressurized on each car, but it no longer serves as a signal pipe to actuate braking. ECP braking systems provide the engineer with data on the status of each car's brake reservoir—its charge and response.

During normal operation or "service braking," ECP brakes can stop trains quicker with reduced bunching. Added precision and control in train braking —made possible by ECP brake's graduated release capability—can improve train handling. Better handling has the potential to improve network efficiency, which would produce fuel savings.[89] Additionally, ECP brakes can make it easier to handle longer trains, which can yield further fuel savings.

ECP braking systems have been around for over 20 years,[90] yet wider adoption has not occurred primarily because ECP systems are expensive and require every car in a train to be equipped with ECP brakes. Safety claims have also been disputed.[91]

Braking with less bunching and better control can also be accomplished with conventional automatic air brakes by locating locomotives in the middle and/or at the end of long trains in a distributed power (DP) arrangement. DP permits the initiation of braking action to begin simultaneously in two or more locations in the train, thus causing cars further back in the train to brake sooner. DP also allows multiple point recharging of the brake pipe. (For additional discussion of distributed power, see Section 4.5).

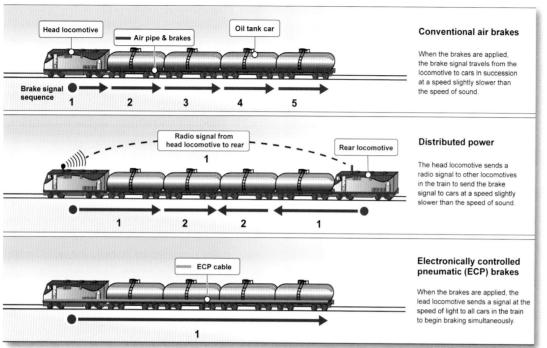

Comparison of conventional train air brakes, distributed power braking, and electronically controlled pneumatic (ECP) brakes from U.S. Government Accountability Office (GAO) *Train Braking* report to Congressional Committees, October 2016. Image Credit: U.S. Government Accountability Office.

Train Brake Energy Efficiency Opportunities. Train brake energy savings can be achieved by using these brakes less—conserving compressed air and ultimately diesel fuel. Compressed air used by train brake systems can also be conserved through maintenance measures that reduce air leakage, which can also occur at connections between railcars. Additionally, ECP brake systems might produce energy savings under some conditions.

Locomotive and Train Parking Brakes

Also called hand brakes, parking brakes are designed to lock locomotives, as well as railcars, in stopped positions. While parking brakes on railcars operate by turning hand-wheels attached to chains and linkages that force friction brakes against wheel treads, parking brakes on all new and rebuilt locomotives are electrically controlled and can be engaged at the push of a button from the cab. When locomotives or cars are parked, their parking brakes should always be applied.

When a train is left unattended, parking brakes should be applied on the locomotive and hand brakes on an appropriate number of cars to secure the train. The number of cars requiring hand brake application is determined by company operating rules, which considers factors such as the weight of the train and the track gradient on which it is parked. A failure to apply a sufficient number of these brakes directly contributed to the catastrophic accident that occurred in Lac-Megantic, Quebec, in 2013.

2.12 Head-End Power

Head-End Power, HEP, or "hotel power" is the power that passenger cars need for lighting, heating and cooling. While electric trains can provide electricity for HEP purposes from the electric power they receive from overhead catenary power lines or third rails, diesel-electric passenger locomotives must generate the power themselves. Three-phase, 480-volt, 60 hertz AC power is needed, and the load may be as much as 600 to 1,000 horsepower (450 to 750 kilowatts). Typically, HEP power is provided by one of these systems:

- A secondary alternator connected to the locomotive's prime mover (e.g., GE P42 Genesis locomotive)
- A dedicated diesel/generator unit, which can operate when the prime mover is shut off (e.g., EMD F59PHI locomotive)
- Regenerative braking (e.g., Siemens SC-44 and EMD F125)

Due to the intermittent and occasional nature of brake applications, regenerative

Runaway oil train explosion in downtown Lac-Megantic, Quebec, on July 6, 2013. Weak safety practices and procedures of the now-defunct Montreal, Maine, and Atlantic Railroad were blamed for this rail disaster that killed forty-seven people. The train was carrying highly volatile Bakken Formation crude oil. Photo Credit: The Canadian Press/Paul Chiasson.

braking's contribution toward HEP in these locomotives is relatively modest compared to the full load.

When this power is provided by the prime mover, it reduces the amount of power that is available for traction purposes. When HEP power is provided by a dedicated unit, savings may be achieved because it makes it unnecessary to idle the locomotive's prime mover at higher rotational speeds. However, whether generated by the prime mover or a dedicated HEP diesel unit, HEP services consume diesel fuel

from the locomotive's fuel tank and thus may be considered a "parasitic load" that reduces fuel economy.

Head-End Power Energy Efficiency Opportunities. There are basically two strategies for reducing HEP energy consumption: (1) Supply Side—Produce HEP more efficiently, and (2) Demand Side—Reduce the amount of HEP that is required in the first place. To accomplish the first, a more energy efficient HEP electrical generator is needed. Another supply-side strategy would be to make

more use of dynamic braking energy to supply power for HEP loads.

HEP loads can be reduced by implementing the same kind of conservation measures that one would employ in a building:

- Reduce heating and cooling loads
  - Better insulate and weather-seal passenger railcars
  - Use the most energy efficient heating, ventilation and air conditioning (HVAC) equipment in each car
  - Experiment with load-shedding strategies that deactivate or cycle HVAC systems for short periods of time while maintaining comfort conditions[92]
  - Use "demand control ventilation" to modulate the outside air supply to passenger car cabins[93]
- Reduce lighting load
  - Install LED lighting
  - Control lighting with motion detectors (e.g., in bathrooms)

Demand control ventilation (DCV) deserves explanation. If a passenger car HVAC system has been designed only to satisfy maximum occupancy conditions, then it always provides a volume of outside air suitable for full occupancy irrespective of actual occupancy or need. The volume of outside air matters because it must be heated or cooled, which consumes energy—in this case, head-end power. Additionally, this ventilation air must be pushed through ductwork by fans, consuming more head-end power. This is a wasteful design because passenger cars are not always full, and air quality goals can often be met at lower outside air supply rates. Fortunately, energy savings can be safely achieved by using DCV, which uses sensors to measure carbon dioxide levels in passenger cars and then uses this data to reduce or increase outside air ventilation rates. Ventilation rates are modulated by controlling dampers and using variable speed drives to control fan speeds and thus air volumes.

2.13 Locomotive Aerodynamics

In the 1930s, lightweight, high-speed, diesel-electric trains were unveiled by the Union Pacific and Burlington railroads, the *M-10000* and *Zephyr*, respectively. With their sleek exteriors designed by industrial designers, these trains were aerodynamic and well suited to 100 mph speeds. These early diesel streamliners competed against very fast streamlined steam locomotives and were soon followed by EMD's iconic E and F type diesel-electric locomotives that gave the appearance of speed while standing still.

Of course, a sleek look can be more about style than aerodynamics, speed and efficiency. We see that in EMD's LWT12 diesel with its car-like appearance. This futuristic design took cues from 1950s GM concept cars.

Air resistance or drag (which refers to frictional force whose value can be expressed in terms of pounds) is a function of the square of the speed or V^2 of an object. And the energy that must be exerted to overcome drag is a function of the cube of the speed or V^3. For these reasons, wind-cheating aerodynamic design is much more important for fast trains than for slow ones.

While not as aerodynamic as high-speed electric locomotives like the Amtrak *Acela*, today's principal U.S. diesel-electric passenger locomotive—the GE P42 Genesis—was designed with function as well as form in mind. The chisel-nosed shape of the Genesis locomotive

Amtrak GE P42 Genesis locomotive in Depew, NY, November 2013. Photo by the author.

An EMD E6 locomotive leading the Santa Fe *Super Chief*, as it speeds through California's orange groves. Image Credit: Author's collection.

may be controversial, but the locomotive itself was considered groundbreaking because of its light weight (for a 4,000-horsepower locomotive), "monocoque" design, and other features.[94] Monocoque design uses the exterior skin to provide structural support, reducing or eliminating the need for a conventional frame.

The newest intercity U.S. diesel-electric passenger locomotives—the Siemens SC-44 Charger and EMD/Progress Rail F125—have aerodynamic features, but do not appear to be as aerodynamic as Motive Power's MP40PH-3C locomotive, which primarily serves the regional rail market.

In contrast, U.S. freight diesel-electric locomotives are all business, even brutish in appearance, with little-or-no attention paid to aerodynamics. To reduce construction costs while improving crashworthiness, their front ends are upright, angular, and boxy. And, as a practical matter, their

Aerodynamic industrial design at the end of the steam locomotive era. Here we see Christopher Ludlow's wonderful painting, "Ego Trip," depicting the 1945, 6,500 hp T-1 4-4-4-4 locomotive and the 1950, 102 hp Studebaker coupe. Their provocative aerodynamic shapes were created by industrial designer Raymond Loewy who is seen driving the Studebaker. Image Credit: Christopher Ludlow (Courtesy of Chuck Blardone).

sides have indented walkways for equipment access. These features create substantial drag whenever locomotives approach their modest top speeds of 70 or 75 mph.

Siemens Charger (SC-44) with standard aerodynamic treatment arriving on Amtrak's Northeast Corridor for testing, August 2016. Photo Credit: Matt Donnelly.

Sleek MP40PH-3C stopped for passengers at Toronto's Union Station, August 2016. Photo by the author.

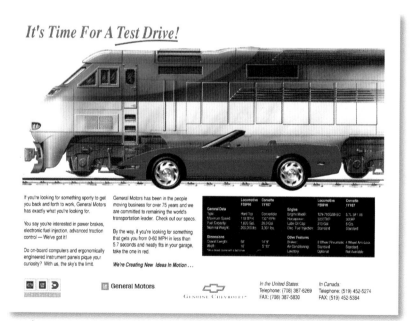

Chevrolet advertisement comparing the aerodynamic good looks of both GM EMD's F59PHI locomotive and Chevy's 1987 Corvette, effectively promoting both. Courtesy of former General Motors Electro-Motive Division, now part of Progress Rail, A Caterpillar Company.

Eye-catching Brightline Siemens Charger with unique nose treatment. Brightline runs between Miami and West Palm Beach (eventually to Orlando) with these 125 mph locomotives on the front and rear of its trainsets. Photo Credit: Brightline.

GM EMD Print Ad, *Trains* magazine, September 1955. Courtesy of former General Motors Electro-Motive Division, now part of Progress Rail, A Caterpillar Company.

EMD LWT12 diesel-led lightweight train. Buffalo, November 1955. Photo Credit: Harold Ahlstrom (Courtesy of the Niagara Frontier Chapter, NRHS).

From an energy perspective, it would seem that the incremental cost of better aerodynamics for line-haul freight locomotives would pay for itself, given the long lifespans of these locomotives. But adding additional cost to an already expensive piece of utilitarian equipment can be a hard sell.

The real aerodynamic issue for diesel-electric freight locomotives is the drag or air friction load generated by the train the locomotive is pulling—especially if it is a long train of irregularly shaped or uneven height cars or, worse, open hoppers that "catch the wind" and can create even more drag in windy conditions. The *Railroad and Locomotive Technology Roadmap* (2002) mentions a report by one railroad stating that fuel consumption for pulling an empty coal train could be equal to that of pulling a full one because of aerodynamic drag[95]—though that report may have neglected to point out that the empties were being hauled uphill into the coal fields. Nonetheless, locomotive engineers describe pulling an empty coal train with, say, 135 cars to be like "pulling 135 open parachutes."

Operating freight trains at higher speed would require more horsepower, resulting in more fuel consumption. It would also lead to more track and equipment maintenance. These are additional costs freight railroads would like to avoid.

Locomotive Aerodynamics Energy Efficiency Opportunities. While new U.S. passenger locomotives have aerodynamic designs, there are still opportunities to improve aerodynamics in order to increase energy efficiency and fuel economy. Freight locomotives are a different story.

Top Speeds of Diesel-Electric Passenger Locomotives

Our discussion of aerodynamics raises the issue of top speed. Just how fast can diesel-electric locomotives go? What are their top speeds?

Presently, both the GE P42 Genesis and EMD F59PHI can be geared for top speeds of 110 mph. However, faster U.S. passenger diesel-electric locomotives are now in service. EMD's 125 mph passenger locomotive, the F125 Spirit, is in revenue service for the Southern California Regional Rail Authority's Metrolink. Siemens 125 mph locomotive, the Siemens SC-44 Charger, is now in use by Brightline, Amtrak, SEPTA and MARC (Maryland Area Regional Commuter). Of course, these locomotives can only reach top speeds if they have suitable track, with high enough track speed limits, little congestion, and enough room between stops to fully stretch their legs.

While 125 mph speeds may be new for U.S. diesel-electric passenger locomotives, this is not the case for British locomotives. British Rail Class 43 diesel-electric locomotives have run at 125 mph for over 40 years as part of British Rail's InterCity 125 High Speed Train service.

Class 43 locomotives operate in pairs, with one power car in the front and one in the rear of the train. While Russian and Spanish diesel-electric locomotives claim to have gone faster, a British Class 43 train is said to hold the *Guinness Book of Records*-certified record for fastest speed achieved by a diesel-electric locomotive—148 mph, reached during a test run in 1987.[96] Brit Rail Class 43 trainsets are now being replaced with Hitachi Class 800 *Super Express* trains. These have operational top speeds of 125 mph but are capable of 140 mph. Significantly, these trains are dual-mode—they have MTU diesel-generator units under the floor of the power units and are capable of operating as electric locomotives when catenary exists (see dual-mode locomotive discussion in Section 6.6).

Electric locomotives and trains have advantages when it comes to top speed, in large part because they have so much power on tap from catenary lines. Amtrak's electric *Acela* trainset reaches speeds of 150 mph along sections of the Northeast Corridor between Boston, Massachusetts, and Washington, D.C. While electric passenger trains operating in Belgium, China, France, Germany, Italy, Japan, South Korea, the Netherlands, Saudi Arabia, Spain, and the United Kingdom have maximum commercial speeds of 186 mph (300 km/hr) to 199 mph (320 km/hr), new Chinese high-speed electric trainsets are operating in commercial service at 217 mph (350 km/hr) and are capable of 249 mph (400 km/hr).[97] (Maglev trains are capable of even higher commercials speeds, though are not part of this discussion.)

History was made on April 3, 2007, when a specially prepared, French Alstom-built electric SNCF TGV (Train á Grande Vitesse) passenger trainset established the world speed record for "conventional wheeled rail vehicles" by reaching an astonishing 357 mph on a test run through the French countryside. The French have been top speed champs for years. Remarkably, as early as 1955 a three-car French passenger train pulled by a 12,000 hp electric locomotive exceeded 200 miles per hour.[98]

In passenger train top speed and commercial speed, the United States lags significantly behind other countries.

Brit Rail Class 43 locomotives lined up at Paddington Station, London, England, October 2011. Photo Credit: Tom Murray, *Trains* magazine.

Above, northbound *Acela* pauses briefly at Wilmington, DE, train station on Amtrak's Northeast Corridor, August 2013. *On the right*, Amtrak "Danger" sign is appropriate because when the 150 mph *Acela* flies by there is not much warning or time to stand back. Photo by author.

Improving their aerodynamics might have limited utility because of their generally slower speeds, but high-speed freights on long-distance routes could benefit by more aerodynamic treatment.

2.14 Other Types of Locomotives

This chapter is about gensets, slugs, and rebuilt older locomotives. These locomotives do not represent all other types of locomotives, but they do have interesting energy stories and are gathered here for brief discussion.

Genset Locomotives

Developed by Union Pacific and National Railway Equipment (NRE) in 2002-2003,[99] genset locomotives are designed to operate in railroad yards and on branch lines.[100] The term *genset* refers to a generator set—a self-contained diesel engine/generator unit based on an advanced low-emissions diesel truck engine. Two or three gensets are installed in a genset locomotive to make up its power generation system.

Genset proponents claim that gensets reduce maintenance time and improve availability because an individual genset can be removed from a genset locomotive and replaced easily if it fails to operate properly. Moreover, the Association of American Railroads and Union Pacific reported that gensets reduce NOx emissions by up to 90% and fuel consumption by up to 37% compared to the locomotives they replace.[101] Railpower, NRE, and Railserve claim energy savings of as much as 55%, 60%, and 72.5%, respectively, for their genset locomotives.[102] The emissions reduction associated with gensets is the result of reduced fuel consumption and the cooler combustion temperatures associated with smaller, truck diesel engines. In theory, genset locomotives can achieve higher levels of overall energy efficiency by turning on and off multiple smaller engines to match loads, allowing the engines that are operating to do so in their energy efficient range. Union Pacific reports that it introduced the first Tier 4-compliant certified genset in 2015.[103]

Union Pacific Railpower genset locomotive with three generator sets. Photo Credit: Mike Danneman.

Not all railroads are satisfied with genset performance. Critics have found these units to be unreliable, slow to respond, lacking sufficient adhesion, and often needing all engines operating to meet load —which would undermine gensets' multi-engine efficiency concept.[104] Critics also maintain that the truck engines in gensets have not been durable enough to absorb the beating they take in locomotive service—leading to breakdowns, additional maintenance, and loss of availability.

Genset Energy Efficiency Opportunities.
Gensets could be more efficient, but are unlikely to see greater use unless effectiveness and reliability issues are addressed.

Slug Locomotives

A *slug* is a heavily ballasted locomotive that has electric traction motors but no prime mover or electric-power-generating alternator of its own.[105] A slug is typically paired with a fully equipped diesel-electric *mother locomotive* whose prime mover and electric generating equipment supply electricity to its own traction motors and to those in the slug. Because tractive force at low speed is limited by adhesion (not horsepower), the addition of the slug results in greater levels of tractive force being applied to the rails. A mother locomotive paired with a slug has almost twice as much low-speed traction as a single locomotive.

The mother-slug pair is also more energy efficient than two regular locomotives because the additional electric demand of the slug's electric motors allows the mother locomotive's diesel engine to operate more often in its higher power, more energy efficient range. Also, when idling is unavoidable, a mother-slug pair saves a considerable amount of fuel because only one diesel engine idles. A slug may be ballasted by filling its fuel tank with concrete or adding steel plate.

Slugs are used to perform low-speed switching but may be seen *on the road* (mainline) providing local service. If operated at more than about 20 mph, the mother locomotive may not have enough horsepower to supply full electrical power to the electric traction motors of both locomotives. For this reason, some mother-slug sets are designed to cut out power to the slug traction motors above 20 mph.

Slugs exist with or without operating cabs. If a cab is present, the mother-slug pair can be operated from either end.

Norfolk Southern has been among the railroads that have built their own slugs,

Kansas City Southern GP22ECO "mother" locomotive with slug, Pueblo, CO, September 3, 2009. Photo Credit: Nathan Zachman.

having set 150 pairs as a goal.[106] Recent Norfolk Southern mother units are built from old EMD GP50 locomotives and use a 3,000 hp GP33ECO EPA Tier 3-compliant engine plus new traction motors, electronics, radiators, cabs and other features.[107] Norfolk Southern states that its mother-slug combinations operate with very low emissions compared to the two switcher locomotives they replace. Typically, Norfolk Southern slugs have full bodies with operator cabs. CSX has also made extensive use of slugs.

Energy Efficiency Opportunities for Slugs.
The energy efficiency of slug locomotives is a function of their in-house

manufacture since railroad companies make their own slugs by cannibalizing their older locomotives. Recycling and reuse conserve material resources, but older components are typically not as efficient as new components. Further energy efficiency improvement is possible by retrofitting slugs with new AC motors and controls—though this costs more. Slugs could also be equipped with batteries, supercapacitors, or other energy storage devices, though their low-speed operation would limit the amount of energy they could recover through regenerative braking.

Repowered Locomotives
Repowering is here defined as a major overhaul that can involve removing everything from an old locomotive's frame and essentially reconstructing it with new or refurbished parts, e.g., prime movers, alternators, traction motors, auxiliaries, cooling systems, comfort cabs, electronics, and microprocessor controls—just about everything!

Class I railroads, with support from EMD and GE, are rebuilding or repowering older locomotives—with some of the more active programs being conducted by BNSF, Canadian Pacific, CSX, and Norfolk Southern.[108] Rebuilt/repowered locomotives can substitute for or postpone the purchase of new locomotives, and, thus, be viewed by Class I railroads as a less expensive way of obtaining modern motive power.[109] Norfolk Southern, the leader in this initiative, has repowered over 450 GP38, GP40, GP50, SD50, and SD60 locomotives in its Altoona, PA, and Roanoke, VA, shops since 2001.[110]

There are multiple reasons for and benefits from rebuilding older locomotives. These include:

- Saving money compared to the cost of new locomotives
- Reducing emissions
- Improving fuel economy
- Reusing and recycling instead of scrapping

Norfolk Southern's Juniata Locomotive Shop floor where repowered GP33ECOs are assembled, Altoona, PA. Locomotive rebuilding is a complicated and expensive undertaking, though the cost of creating a newly rebuilt locomotive may be half that of purchasing a new locomotive. Photo Credit: © Norfolk Southern Corp.

Progress Rail's EMD reports that their 710 ECO repowered locomotives require only 18 tons of new steel while a comparable new locomotive would have required 111 tons of steel to construct—for a saving of 93 tons of steel.[111] Assuming that the manufacture of 1 ton of steel requires 2,500 pounds of iron ore, 1,000 pounds of coal, 40 pounds of limestone, and 75,000 gallons of water,[112] then each repowered EMD 710ECO produces these material savings:

- 232,500 pounds (116 tons) of iron ore
- 93,000 pounds (46.5 tons) of coal
- 3,720 pounds (1.9 tons) of limestone
- 6,975,000 gallons of water

If a rebuilt locomotive is going to be used for yard or local work, the locomotive that began as a DC locomotive may continue to use DC traction motors. However, some rebuilt DC road locomotives have been converted to AC motors, e.g., those produced by GE's "DC2AC" program.[113] This program has converted BNSF and Norfolk Southern GE DC Dash 9 locomotives into AC locomotives with tractive efforts equal to the latest new AC locomotives—200,000 pounds of starting tractive effort with an adhesion factor of 46%.

A common rebuild uses EMD 710ECO kits that provide new 8-cylinder (2,150 hp) or 12-cylinder (3,150 hp) EMD 710

diesel engines, which are highly regarded for durability and longevity. The number 710 refers to the displacement in cubic inches of one cylinder of this engine. Also included in the kit is a new alternator, aftercooling loop, AESS (Automatic Engine Start Stop), and microprocessor system.[114] Thus, these kits provide a near-complete rebuild.

According to EMD, their ECO repowering kits reduce nitrogen oxides, particulate emissions, unburnt hydrocarbons, and carbon monoxide by 50% to 70%, produce

fuel savings of 20% to 40%, and increase adhesion by 15%—compared to the performance of old (un-repowered) locomotives.[116]

In early 2017, Progress Rail announced that it had successfully completed emissions testing of a new locomotive, the 2,000 hp EMD24B Repower-T4—a Tier 4-compliant switching locomotive built using a remanufactured GP-40 frame and cab.[117] An alternative to Tier 4 gensets, the EMD24B's first assignment is working at the ports of Los Angeles and Long Beach, CA, as part of the Pacific Harbor Line's

A unique rebuild. TP56 switcher, designed by Traction Power and built by Curry Rail Services, is made by cutting an SD40-2 frame in half.[115] Photo by the author.

fleet of low-emissions locomotives. The locomotive uses a urea SCR catalyst system to reduce NOx emissions to T-4 levels (see Section 5.3).

General Electric also provides locomotive repowering services and kits. GE's program replaces locomotive operator, electrical, and radiator cabs. It also modernizes control and propulsion systems and installs GE's Trip Optimizer, distributed power system, and PTC equipment. Claimed benefits include up to 50% more tractive effort, 25% increased adhesion, 10% greater fuel economy, 40% improved reliability, and 20% reduction in maintenance and repair costs.[118]

Energy Efficiency Opportunities When Repowering Locomotives. Repower programs can push the energy efficiency envelope by installing the most energy efficient technology available at the time of the retrofit. Repowering also presents an opportunity to retrofit waste heat recovery systems (see Section 6.2). Since refurbished equipment can be less energy efficient than new equipment, quality control is essential when using rewound alternators and motors, for example.

De-Rated Locomotives

CSX determined that it could save energy by creating a small fleet of de-powered existing locomotives. From 1999 to 2006

Norfolk Southern GP33ECO repowered locomotives line up at Inman Yard, Atlanta, GA. The locomotives were formerly GP50s. Photo credit: © Norfolk Southern Corp.

CSX undertook a program that permanently reduced the horsepower of 132 SD60 and SD70 locomotives by 15% to 25%.[119] This reduction enabled these locomotives to operate at the needed lower horsepower outputs while in more energy efficient throttle settings. Fuel savings were estimated to be 10% to 15% with reduced emissions and maintenance requirements. For the same reason, some Norfolk Southern diesel-electric road locomotives have switches to reduce

nameplate 4,400 traction horsepower to 4,000 horsepower. Maximum horsepower is dialed back by maintenance staff in the mechanical department, not the engineer. The loss of 400 horsepower slightly affected acceleration, but otherwise was operationally acceptable while producing the benefits mentioned above.[120]

Summerhill, PA, June 2015. Photo by author.

CHAPTER THREE

Diesel-Electric Locomotive Fuel Economy and Energy Efficiency

Let's first take a look at the energy efficiency advantages of trains over other modes of transportation, and then the various measures or metrics of railroad and locomotive energy efficiency. "Energy efficiency" and "fuel economy" are used here interchangeably, most of the time, though the former term is more general and encompassing since it does not only refer to diesel fuel consumption. The absence of public information from locomotive manufacturers about the efficiency of their products will force us to do some speculating and educated guessing on this topic.

3.1 Energy Advantages of Railroads and Trains

Railroads have built-in energy efficiency advantages over other forms of transport. That edge is derived from economies of scale and the ability of a train to move with minimal mechanical and aerodynamic friction.

Rolling Resistance Advantage

Trains have minimal rolling resistance because they have steel wheels that roll on steel rails. Steel is a very hard material and there is minimal energy-absorbing deformation of either the wheels or rails when a train rolls over a well-maintained straight or "tangent" track.

Friction is further reduced in rail transport by the very small contact patch between a locomotive wheel and rail—often described as the size of a dime or less than 0.4 square inches. Per ton of freight hauled, truck tire contact areas are a thousand times larger.

While deflecting rail, track ties, and ballast absorb some energy and friction

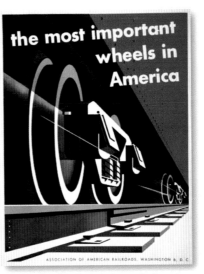

Image Credit: Association of American Railroads.

increases on poorly maintained track or curved track (where wheels partially skid around turns and wheel flanges push against the inside surface of the outside rails), the efficiency of steel on steel is still much greater than competing means of transport. There is no comparison between the relatively small amount of friction generated by steel wheels rolling on steel rails and rubber tires rolling on asphalt or concrete roads.

Of course, locomotives need friction when starting, accelerating, climbing hills, and braking. Thus, reduced friction is not always a virtue, but on balance it has served railroads well.

Aerodynamic Advantage

Trains also have an inherent aerodynamic advantage because they are long and narrow and have a very small cross section per unit of cargo or per number of cars or passengers. Essentially, locomotives punch a hole in the air and following cars draft behind the lead locomotive. This produces proportionately less air resistance than trucks, buses, or cars. This is true even of trains pulled by not-so-aerodynamic freight locomotives.

The next sections of this chapter define railroad and locomotive energy efficiency or fuel economy in terms of either work output per unit of energy input or energy consumed per unit of work output. There are multiple definitions, and they lead to confusion. This chapter attempts to carefully delineate these definitions and make them understandable. The author wishes this important part of the book was less complicated.

3.2 Railroad Corporate Average Fuel Economy

This type of fuel economy is not a measure of locomotive energy efficiency but is presented here because it is the only type of railroad fuel economy that is widely publicized. Railroad corporate average fuel economy has to do with how much "stuff" the railroad industry or individual railroads can move on a given amount of fuel—typically per gallon. It is a measure of annual systemwide hauling activity versus annual systemwide fuel

consumption. As such, it is affected by locomotive fuel economy—the design efficiency of locomotives and how efficiently they are operated—but other factors contribute. They include total annual hauling, train load management (e.g., the amount of freight or passengers per car and per train), rail system traffic management, fuel consumed in yard activities, and whether a particular railroad runs on largely flat territory (e.g., Canadian National) or predominantly mountainous territory (e.g., Canadian Pacific).

There are at least three versions of railroad industry corporate average fuel economy:

Association of American Railroads Revenue Ton-Miles per Gallon

The best-known version of railroad corporate average fuel economy is annually generated by the Association of American Railroads. It is calculated by dividing the annual revenue-producing "ton-miles" of freight of all the seven U.S. Class I freight railroads by the number of gallons of diesel fuel all these companies consumed that year. A "ton-mile" is just what it sounds like, i.e., moving a ton 1 mile. A "revenue ton-mile" is equal to moving a ton of revenue-producing goods 1 mile. Revenue tons are determined by subtracting the *light*

AAR's myth-buster print ad from 1979 pointing out that railroads are in the forefront of energy conservation. Since then, the industry's fuel economy has doubled, as shown by the bar chart. Image Credit: Association of American Railroads.

or *tare weight* of the freight cars (their empty weight) from their total loaded weight. With this industry-wide metric, a *higher* number reflects improved fuel economy. Similarities with automobile efficiency are evident given that this fuel economy measure is also in *miles per gallon*—miles per gallon for moving 1 ton of revenue producing freight.

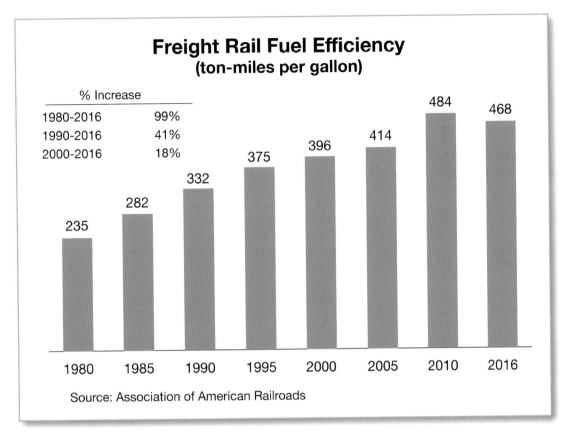

Freight Rail Fuel Efficiency
(ton-miles per gallon)

% Increase	
1980-2016	99%
1990-2016	41%
2000-2016	18%

1980	1985	1990	1995	2000	2005	2010	2016
235	282	332	375	396	414	484	468

Source: Association of American Railroads

(GTM) per gallon metric and compare it to other railroads. This measure provides an indicator of how much overall weight (cargo plus the weight of the trains themselves) they move per gallon of fuel consumed. The GTM/gallon metric also captures the weight and mileage a railroad undertakes to move ballast, ties, and rail in work trains. A large rail network runs dozens of work trains daily. The revenue ton-mile (RTM)/gallon measure is used less often by individual railroads. If there is a big discrepancy between the two measures, it may suggest moving a lot of empty cars around the network. Some railroads flip these measures to gallons/GTM and gallons/RTM. All railroad company internal efficiency rating systems are influenced by commodity volumes, freight mixes, routes, and traffic management.

As of 2016, U.S. Class I railroads could move, on average, a ton of revenue-producing freight 468 miles on a single gallon of diesel fuel—which is pretty amazing.[121] With good reason, the AAR and the railroads proudly publicize this fact. This measure of fuel efficiency shows a decline since 2010 when a ton of revenue-producing freight was transported 484 miles on a single gallon of diesel fuel. Hopefully, this decline is temporary. Among other things, it is due to reduced rail shipping volumes. As a general rule, higher shipping volumes enable higher RTMs/gallon. Worthy of note is a 2009 Federal Railroad Administration study which concluded that freight railroads were nearly four times as efficient as the trucking industry.[122]

Individual Railroad Company GTM and RTM per Gallon
Typically, individual railroad companies compute their own 1,000 gross ton-miles

Amtrak BTUs per Passenger-Mile
This metric changes the work units to "passenger-miles" and energy units to BTUs or British Thermal Units. It also flips the equation, making energy consumption (BTUs) the numerator and revenue-producing cargo movement (the passenger-miles) the denominator. A *passenger-mile* is also what it sounds like, i.e., moving one passenger 1 mile. A BTU, as previously described, is equal to the amount of heat it takes to raise the temperature of a pound of water 1°F.

89

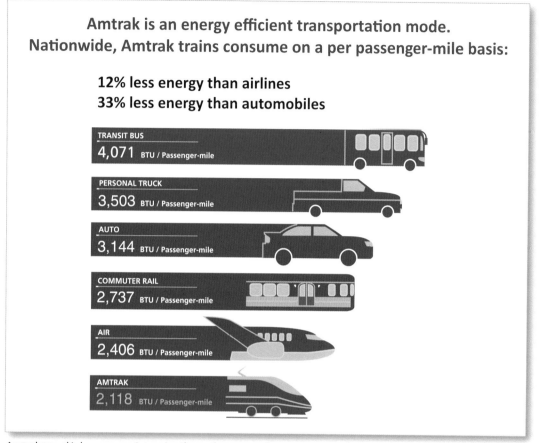

Amtrak is an energy efficient transportation mode. Nationwide, Amtrak trains consume on a per passenger-mile basis:

12% less energy than airlines
33% less energy than automobiles

TRANSIT BUS
4,071 BTU / Passenger-mile

PERSONAL TRUCK
3,503 BTU / Passenger-mile

AUTO
3,144 BTU / Passenger-mile

COMMUTER RAIL
2,737 BTU / Passenger-mile

AIR
2,406 BTU / Passenger-mile

AMTRAK
2,118 BTU / Passenger-mile

Amtrak travel is less energy-intensive than other types of passenger transportation. Source: Transportation Energy Data Book, 2015. Image Credit: Amtrak.

Like the freight railroad measures above, Amtrak's BTUs per passenger-mile metric uses annual systemwide data. With BTUs per passenger-mile, a *lower* rating represents better fuel economy. A BTU-based measure allows the use of the different fuels (diesel fuel, gasoline, jet fuel, and electricity) to be compared on a common basis.

3.3 Locomotive Fuel Economy

There are at least four different measures of diesel-electric locomotive fuel economy, and, *none of them* provides a measure of the energy efficiency of the *entire* locomotive. Rather, they provide fuel economy for a portion of the loco-

motive. While this discussion seeks to explain and precisely define measures of locomotive fuel economy, in reality this subject is more complicated than presented here, and the definitions have been less than ideal, allowing locomotive manufacturers the latitude to self-define them.

The four measures of fuel economy are:

- Net Traction Specific Fuel Consumption (NTSFC)
- Brake Specific Fuel Consumption (BSFC)
- Brake Horsepower-Hours per Gallon (BHP-hr/gal)
- Fuel Consumption Rate for Different Throttle Settings

While four measures of locomotive fuel economy are presented here, the railroads are primarily interested in the first two, i.e., NTSFC and BSFC. On the other hand, regulators such as the Environmental Protection Agency (EPA) and California Air Resources Board (CARB) are interested in the BHP-hr/gal measure. While gallons of fuel consumed per hour for different throttle settings is not, strictly speaking, a measure of fuel economy or energy efficiency, railroads and locomotive leasing companies often use this type of data as a kind of shorthand for fuel economy.

Fuel Use vs. Fuel Economy

This book is premised on the importance of improving locomotive efficiency. However, a larger concern from both environmental and economic perspectives is the overall amount of diesel fuel that a locomotive consumes. This is affected by all these factors:

- Locomotive age, model, and design efficiency
- Locomotive operation
  - Engineer training, motivation, and skill
  - Locomotive energy management systems
  - Train speed
  - Train handling
  - Idling
- Fuel type
- Train length and weight
- Train makeup and aerodynamics
- Train car maintenance
  - Condition of wheels and bearings
- Route length, gradients, and curvatures
- Rail network management
  - Rail system traffic volume
  - Congestion
  - Train routing
  - Two-way hauls
  - Track work en route
- Roadway and track condition
  - Maintenance
  - Track elasticity
  - Wheel/rail interface
  - Wayside lubrication
- Fleet maintenance
  - Truck and bearing maintenance
- Weather
- Season

Different values for the factors listed can make the difference between trains that move freight at the rate of 200 ton-miles per gallon versus 1,000 ton-miles per gallon.[123] These factors also make it difficult to accurately compare the efficiencies of locomotives on the road.

Railroads can control some of the factors listed but not all of them, e.g., weather and season. Head winds, tail winds, and cross winds affect fuel consumption rates and overall use. Winter operation increases fuel use because locomotives idle more (to keep cooling water from freezing), axle bearings on locomotives and train cars take longer to warm up, and train brakes stick on long grades and take longer to fully release at the rear of the train.

Flat spots on car wheels adversely affect the wheel/rail interface, increasing rolling friction and fuel consumption. The installation of Wheel Impact Load Detectors (WILD), which identify car wheel flat spots, has the potential to improve fuel efficiency. Track elasticity or "modulus" affects fuel consumption as well. Properly maintained roadway with stiff track—supported by well-maintained roadway—absorbs less energy than track that deflects significantly when weight is applied to the rails.

Train aerodynamics is an underdeveloped opportunity that freight railroads need to investigate further to adequately compete with other transportation modes. Attempts have been made by Union Pacific in the past with an intermodal fairing called the ArroWedge and a complete train set of aerodynamically modified covered hoppers that have produced documented 4 to 6% fuel savings.

Types of Locomotive Horsepower

Because NTSFC and BSFC are defined in terms of different types of locomotive horsepower—i.e., net traction, traction, and brake horsepower—those types of horsepower will be defined first. Note that NTSFC and BSFC fuel economy ratings, and the horsepower types that define them, reflect steady-state, self-loading, stationary testing under AAR test conditions.[124] These conditions do not necessarily align with real world operation.

Net Traction Horsepower (NTHP). We

start with this type of horsepower because it can be measured and proven on the locomotive. NTHP is the power measured at the output of traction alternator and rectifiers that serve the traction motors. DC volts and amps are actually measured. Their product equals kilowatts or megawatts, which can be converted into horsepower. Of course, NTHP, like traction horsepower and brake horsepower, varies by throttle setting.

The advent of Tier 4 locomotives slightly changes the way NTHP is described because Tier 4 locomotives have only one alternator that serves both traction and auxiliaries' functions. Accordingly, the NTHP of a Tier 4 locomotive is that portion of the power output of the (single) alternator's rectifiers that serves the traction motors.

Traction Horsepower (THP). THP is

the horsepower delivered to the traction alternator or, in the case of Tier 4, to the single alternator for traction purposes. As such, traction horsepower is a calculated value. THP is calculated by dividing NTHP by the efficiency of the traction alternator and rectifiers. Since the efficiency of the traction alternator and rectifiers is not measured but rather is provided by the locomotive manufacturers, THP is a theoretical value from the railroad's perspective. In the case of Tier 4 locomotives, that efficiency would be for the single alternator and traction rectifiers.

Brake Horsepower (BHP). BHP, some-

times called gross horsepower, is the prime mover's shaft output. It is the horsepower output of the engine itself. It serves the locomotive auxiliaries (including shaft-driven auxiliaries in the case of many older EMD locomotives) and alternators.

BHP is a calculated value because shaft horsepower cannot be measured directly in a locomotive. BHP is calculated by subtracting the total horsepower of the auxiliaries from the THP. The total horsepower of the auxiliaries may be provided by the manufacturers or measured by the railroads. However, auxiliary loads used in these calculations traditionally do not include operation of such necessary devices as lighting, PTC, and voice radio (approx. 6 hp); cab HVAC (approx. 9 hp); and the air compressor (approx.

55 hp). Thus, BHP is also a theoretical value from the railroad's point of view.

Measures of Locomotive Fuel Economy

Now that NTHP and BHP have been defined, we can define NTSFC and BSFC measures of locomotive fuel economy, as well as the other two measures, i.e., BHP-hr/gal and rate of fuel consumption for different throttle settings.

Net Traction Specific Fuel Consumption.

NTSFC is defined as the number of pounds of diesel fuel consumed per hour divided by the NTHP produced. As such, it provides measurable fuel economy or energy efficiency of the locomotive up to but not including the traction motors, gearing, axle (journal) bearings, and wheel/rail interface. While not drawbar efficiency, it is nearly full locomotive efficiency. Because this measure of fuel economy is based on dividing fuel use by power output, a *lower* NTSFC value represents higher fuel economy. NTSFC may be provided for each throttle setting or notch.

Brake Specific Fuel Consumption.

BSFC is the number of pounds of diesel fuel consumed per hour divided by the BHP produced. As such, BSFC represents the fuel economy of the prime mover only. In locomotives this rating of fuel economy is theoretical because it's based on BHP which is, as we have seen, based on assumptions. As in the case of NTSFC, a *lower* BSFC rating represents higher fuel economy, and BSFC may be provided for each notch.

Brake Horsepower-Hours per Gallon.
BHP-hr/gal flips the equation because it is a measure of energy or work output divided by energy input. BHP-hr/gal is the number of horsepower-hours of work output of the prime mover per gallon of diesel fuel consumed. Here again, this rating is theoretical in locomotives because their BHP is based on assumptions. The unit horsepower-hour may be unfamiliar, but it's easy to understand: 1 hp-hr is the energy or work output of a 1 horsepower motor running at full output for 1 hour. A *higher* value means better fuel economy with this measure of fuel economy.

Norfolk Southern GE C44-9W locomotive leading intermodal up the famous Horseshoe Curve, Altoona, PA. January 2018. Photo by author.

Consider this test data from a GE Dash 9 (C44-9W) locomotive. Ranging from 14–25 years old, thousands of Dash 9s remain in service as a mainstay of many Class I railroads. Here we see that Notches 7 and 8 have the best fuel economy (the lowest values). Note also that BSFC values for any notch are lower (representing higher efficiency) than NTSFC. That makes sense because NTSFC includes the additional energy consumption of the auxiliaries and alternator/rectifier.

Notch	BHP	Fuel	BSFC	NTHP	NTSFC
		lbs/hr	(lbs/hr)/BHP		(lbs/hr)/NTHP
8	4500	1500	0.33	4250	0.35
7	3675	1215	0.33	3500	0.35
6	2975	1010	0.34	2775	0.36
5	2230	800	0.36	2065	0.39
4	1575	575	0.37	1445	0.40
3	1050	400	0.38	975	0.41
2	525	225	0.43	460	0.49
1	210	100	0.48	180	0.56

As a general rule, a BHP-hr/gal rating is based on the EPA duty cycle that allocates percentages of time that locomotives theoretically spend, on average, in each notch setting. The EPA duty cycle for line-haul locomotives is shown below. The EPA uses a different duty cycle for switching locomotives.

The EPA used this "duty cycle" when calculating a fuel economy of 20.8 BHP-hr/gal for line-haul diesel-electric locomotives in actual operation.[125] This average, also used by CARB, does not discriminate by locomotive type or age. It is used in conjunction with emissions regulation.

This data serves an important consciousness-raising purpose by telling engi-

Fuel Consumption Rate for Different Throttle Settings.

To promote fuel economy, some railroads provide their engineers with information about how much fuel per hour their locomotives burn in various throttle settings. For example, one Class I railroad presented this data to its engineers:

Fuel Burn by Notch (gallons per hour)

	THP	N8	N7	N6	N5	N4	N3	N2	N1
GE ES44AC	4,380	210	171	140	109	79	53	27	12
EMD SD70ACe	4,000	187	164	133	86	64	47	23	12

neers that their locomotives consume a lot of fuel, and increasingly so as higher notches are used, but it does not actually convey fuel economy or efficiency. We can, however, use this data to calculate fuel economy in the form of work output per energy input using simple arithmetic. For example, at full power (Notch 8), the GE ES44AC locomotive described in this chart produces 4,380 THP-hr of

Throttle Setting	Run Time
Low Idle	19.0%
Normal Idle	19.0%
Dynamic Brake	12.5%
Notch 1	6.5%
Notch 2	6.5%
Notch 3	5.2%
Notch 4	4.4%
Notch 5	3.8%
Notch 6	3.9%
Notch 7	3.0%
Notch 8	16.2%

BNSF EMD SD70ACe awaits its next assignment. Buffalo, NY, March 2014. Photo by author.

work every hour. If we divide that value by the amount of fuel consumed to produce that work, i.e., 210 gallons, we get 20.9 THP-hr/gal, which is a measure of fuel economy. The SD70ACe locomotive in this example achieves 21.4 THP-hr/gal. The same type of calculation could be performed for Notches 1 through 7 if we had THP values for those notches.

Fuel Economy by Manufacturer

It is difficult to tell with any certainty which manufacturer produces more energy efficient locomotives. While GE advertises that its locomotives are more energy efficient than EMD locomotives,[126] it does not provide any public documentation to back up that claim. Moreover, static test conditions and manufacturer-provided information can bias comparative fuel economy comparisons and may not be reflective of fuel economy in actual operation. While GE and EMD locomotives have performance differences, for our purposes we will assume that comparable EMD and GE products have comparable fuel economy.

Fuel Economy and the Purchasing Process

Railroads may request that locomotive manufacturers provide fuel economy data in a variety of ways during the purchasing process. NTSFC is of much greater interest to prospective buyers because NTSFC is measurable and verifiable. Railroads may request NTSFC fuel

economy performance specifications for select throttle settings or notches, or for all notches. It is not uncommon for railroads to insist on "fuel guarantees." These are potential financial penalties imposed on the builders if their products do not perform to fuel economy specs based on the agreed upon post-delivery testing. As explained in the following sidebar, this fuel economy information is generally confidential, protected by non-disclosure agreements.

When comparing the fuel economy of one locomotive builder's products to the other's, lower fuel economy is not necessarily disqualifying. The railroads conduct net present value analyses of competing proposals and add cost to the bidder whose locomotives are less efficient. That additional cost is based on the number of additional gallons of diesel fuel that the less efficient locomotive is projected to consume over the "life of the asset" (its anticipated lifespan). Typically, this is a sophisticated calculation based on the railroad's duty cycle (the number of hours per year a new locomotive is expected to operate in each notch) and diesel fuel price projections. Thus, while superior fuel economy and lower life cycle fuel costs contribute positively toward a purchasing decision, they may not be the deciding factors. All costs are considered and they are weighed against the purchase price. The goal of the purchasing

process is not necessarily to buy the most efficient locomotive but to obtain the highest net present value or profit from this sizable capital investment.

TEST RECORDS OF NEW "SUPER" LOCOMOTIVE RELEASED!

New Diesel-Electric already in service hauls more, further, faster, at lower cost!

American Locomotive

A GREAT NEW DIESEL-ELECTRIC FOR EVERY RAILROAD JOB

Early secrecy about diesel-electric locomotive performance reflected in 1946 American Locomotive Company (ALCO) print advertisement. This ad announced a new "revolutionary, super locomotive," "disguised for test purposes," embodying "secret, war-born developments." The locomotive in question was ALCO-GE's 1,500 hp FA diesel-electric freight locomotive. ALCO produced 75,000 locomotives. Most were steam locomotives including the gigantic 4-8-8-4 Union Pacific *Big Boys*. As a diesel-electric locomotive builder, ALCO had trouble competing against EMD and GE, and closed its doors in 1969. Image Credit: ALCO/Eaton.

Confidentiality About Locomotive Fuel Economy and Efficiency

NTSFC and BSFC are standard measures of fuel economy used by locomotive manufacturers to specify locomotive fuel economy. But this data is not public information. Locomotive fuel economy specifications are regarded as trade secrets—labeled "Confidential Business Information"—by the manufacturers. The confidentiality of this information is enforced by "NDAs" or non-disclosure agreements with the railroads. But why the secrecy? Here are some possible reasons or rationales:

- Locomotive manufacturing is a competitive business, and it is natural for GE and EMD/Progress Rail to want to keep product information from each other in order to maintain a competitive advantage during complex sales negotiations. However, while confidentiality agreements slow down the transfer of information about locomotive energy efficiency, sooner or later these manufacturers acquire each other's information.

- The practice of keeping locomotive efficiency data confidential may find railroad industry support because it potentially serves the interests of industry insiders. The "insiders" referred to here are railroad purchasing teams, who are skilled in using confidential data to potentially strike better deals, and those experts in locomotive testing facilities, whose knowledge of confidential data enhances their roles as industry resources.

- Locomotive efficiency is so complex and difficult to understand that the industry may prefer it remain confidential to avoid confusion or the need to explain it. Of course, if the concern here is that public misunderstandings could result when specified (optimal) efficiencies are not achieved in actual operation, then this issue could be addressed by offering a strong disclaimer, e.g., "NTSFC and BSFC fuel economy specifications represent peak design efficiency achieved only under optimal specified test conditions. Fuel economy in actual use is a function of numerous operational variables and environmental conditions and will be lower."

- Finally, locomotive efficiency data is probably confidential because it can be. No one is clamoring for transparency and there is no federal requirement that efficiency data be made public, unlike for other types of equipment.

The bottom line: A single locomotive can consume more than half a million dollars' worth of diesel fuel annually, yet its fuel efficiency specifications are unpublished and unavailable. Even though understanding the nuances of locomotive fuel economy can be difficult, public disclosure of the data—and the testing procedures used to generate it—might still benefit analysts, researchers, and other interested parties including the railroads themselves. If locomotive fuel economy specifications and testing procedures were public knowledge, it probably would spur increased competition to produce more efficient locomotives.

3.4 Locomotive Design Energy Efficiency

Finally, we arrive at "fuel-to-rails" efficiency of the entire locomotive under optimal design or test conditions, expressed as a percentage. What is the "state of the art" in locomotive design efficiency? Given that the newest Tier 4 locomotives are probably not as energy efficient as Tier 3 locomotives (see Section 5.4), the best-case efficiency is likely Tier 3 locomotive technology.

As just explained, locomotive manufacturers do not disclose locomotive fuel economy information. They also do not provide design fuel-to-rails energy efficiency data. Fortunately, we can estimate this whole-locomotive efficiency ourselves in a two-step process. The first step is to identify the peak efficiencies of all locomotive drive train components. These components and their estimated efficiencies are shown on the block diagram to the right.

Note that the diesel engine efficiency is shown as a range—from 42% efficiency in the *Railroad and Locomotive Technology Roadmap* to 48%, which the author understands is the best locomotive prime mover design energy efficiency at this time,[127] assuming the lower heating value of diesel fuel. As a point of reference, 4,000 hp EMD marine diesel engines (for which measured BSFC data are publicly available) have peak efficiencies around 44%.

The second step in calculating overall design fuel-to-rails efficiency is to multiply all the individual component efficiencies together. Here we will use the highest locomotive prime mover efficiency in the range above, i.e., 48%.

Diesel engine at the crankshaft	0.480
Deducted for 200 hp auxiliary load	0.955
Alternator/rectifier	0.965
Inverters	0.990
Motors	0.960
Gearing and journal (axle) friction	0.990

0.480 x 0.955 x 0.965 x 0.990 x 0.960 x 0.990 = 0.416 or ~42%

Thus, we see that the state-of-the-art diesel-electric locomotive fuel-to-rails design efficiency is approximately 42%—an admittedly high number. This estimate assumes ideal conditions and that all drive train devices would be able to achieve peak efficiency at the same time. It also assumes that auxiliary loads are not maximum values.[128] Additionally, potential energy penalties associated with locomotive air resistance and wheel-to-rail rolling resistance are not included. Note that the same fuel-to-rails calculation would yield 0.364 or ~36% if the prime mover's efficiency was 42%—at the lower end of our 42% to 48% prime mover efficiency range.

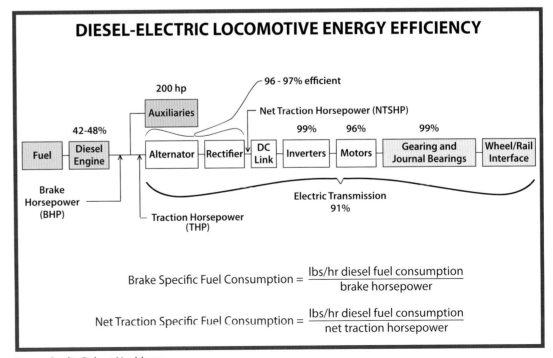

Image Credit: Robert Hochberg.

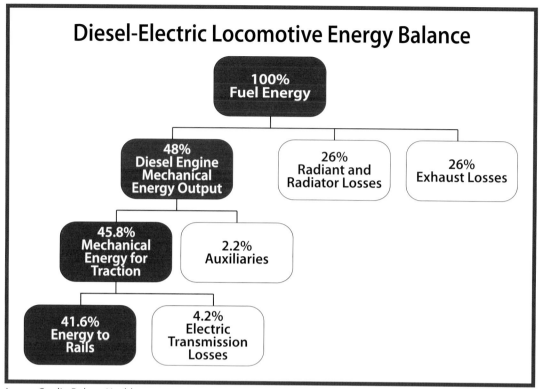

Diesel-Electric Locomotive Energy Balance

100%
Fuel Energy

48%
Diesel Engine
Mechanical
Energy Output

26%
Radiant and
Radiator Losses

26%
Exhaust Losses

45.8%
Mechanical
Energy for
Traction

2.2%
Auxiliaries

41.6%
Energy to
Rails

4.2%
Electric
Transmission
Losses

Image Credit: Robert Hochberg.

A locomotive may excel in ideal or AAR stationary test conditions yet perform less impressively in actual operation, perhaps especially when at higher altitudes or warmer ambient temperatures.

While stationary testing measures fuel economy based on constant horsepower output for each throttle notch, in real world operation notch horsepower varies. At all speeds, and especially on slippery rail (which exists even in dry conditions), traction control systems drive the locomotive's wheels to slip slightly to gain maximum adhesion (see discussion of creep control in Section 2.9). These continuous, fine adjustments in horsepower affect fuel consumption and efficiency. Operational efficiency is also affected by the use of energy management technology and by engineer interest and skill.

An energy balance or flow diagram provides another way of visualizing the efficiency and inefficiencies of the locomotive. The diagram below assumes the 48% prime mover design efficiency, and "electric transmission losses" in the diagram refer to energy losses associated with that segment of the drive train bounded by the electric alternator on one side and the wheels on the other.

While 41.6% or (approximately) 42% overall locomotive design efficiency is excellent,

there is still much energy waste—58% in fact. This inefficiency materializes as waste heat rejected into the environment. It represents opportunity for conservation and efficiency improvement.

3.5 Locomotive Operational Efficiency

In many ways, locomotive operational efficiency is very different from locomotive NTSFC and BSFC fuel economy specifications and peak design locomotive efficiency.

Additionally, we have seen in the sidebar in Section 3.2 that there are many factors affecting operational fuel use. These can adversely impact operating fuel efficiency. Thus, it's important not to confuse the fuel economy or energy efficiency that is measured or calculated under ideal test conditions with the fuel economy that locomotives achieve during normal operation.

These observations have led some to suggest that competing locomotives need only achieve design or test efficiencies within a percentage point or two of each other—within what might be called a *zone of indifference*—to be considered as operating in rough fuel economy parity. It has been argued that GE and EMD locomotives operate within this zone.

As previously mentioned, the EPA and CARB use an average "duty cycle" fuel economy factor of 20.8 BHP-hr/gallon when estimating the efficiency line-haul diesel-electric locomotives in actual operation.[129] In a recent report to CARB, RAILTEC assumed that the average fuel-to-rail efficiency of diesel-electric locomotives in normal operation in California was 33%[130] —quite a few percentage points lower than our design efficiency.

3.6 Prospects for Diesel-Electric Locomotive Efficiency Improvement

Railroads can accelerate the process of developing more energy efficient locomotives by working together to produce a genuine "next generation" energy efficiency specification for the locomotive manufacturers. In that way, the railroads can be more proactive. They can take control and directly challenge the manufacturers to build more energy efficient locomotives, instead of accepting the efficiency of the products the manufacturers make available.

Consider this cautionary tale: When the U.S. Department of Transportation administered the "Next Generation Corridor Equipment"

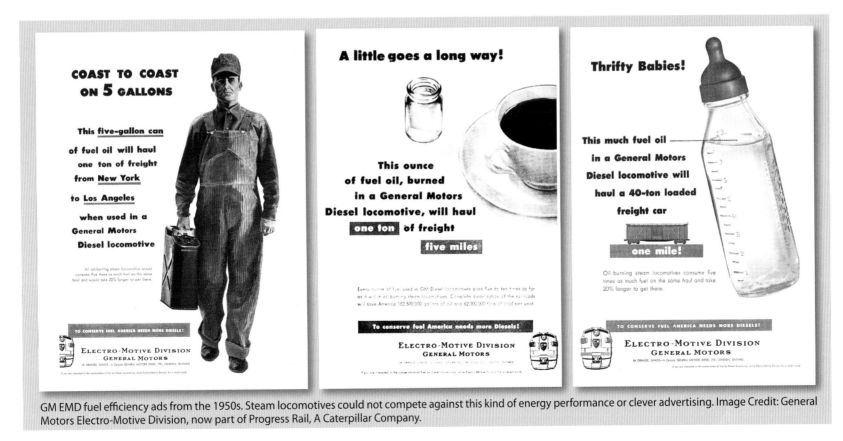

GM EMD fuel efficiency ads from the 1950s. Steam locomotives could not compete against this kind of energy performance or clever advertising. Image Credit: General Motors Electro-Motive Division, now part of Progress Rail, A Caterpillar Company.

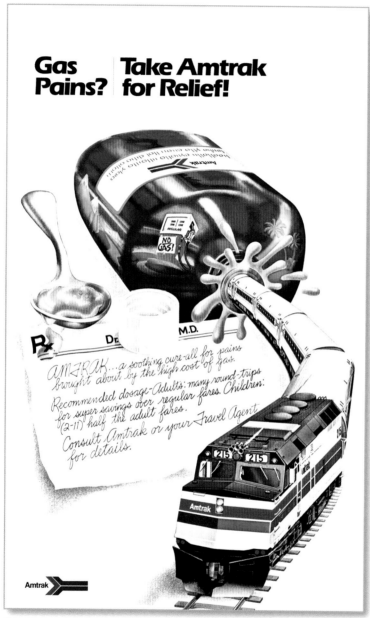

Humorous Amtrak print ad published very briefly during the 1970's energy crisis when gasoline prices skyrocketed. Taking the train did relieve gas (cost) pains. Image Credit: Amtrak.

selection process mandated by the Passenger Rail Investment and Improvement Act (PRIIA) of 2008, it could have developed challenging energy efficiency requirements for new passenger locomotives, but apparently it did not. Here is the efficiency language in the PRIIA specification:

- The locomotive should be "high fuel efficiency, in particular with regard to specific passenger service requirements."
- The locomotive should be capable of providing "efficient performance in intercity rail services in North America."
- "Fuel consumption shall be optimized for the lowest possible life cycle costs."[131]

This sounds good as far as it goes, but there is no specificity. There is no pushing the envelope. While the selected locomotive (the Siemens Charger) is undoubtedly a state-of-the-art energy efficient locomotive, it's possible that stronger, much more detailed energy efficiency language could have improved the state of the art. Of course, it's possible that more detailed assertive energy efficiency specifications were developed by the PRIIA process but treated as confidential. But this was a taxpayer-funded undertaking where transparency should have been the rule.

The potential for very substantial energy efficiency improvement was recognized by railroad experts and government scientists who met together in 2002 to produce the aforementioned *Railroad and Locomotive Technology Roadmap*.[132] This report established an industry-wide goal of improving railroad fuel efficiency 25% by 2010 and 50% by 2020, compared to their 2002 baseline. Here, the measure of efficiency was not locomotive fuel economy, but rather the annual AAR industry measure of RTM/gallon. To achieve these goals, it was expected that energy efficiency improvements would be implemented in many areas of railroad operation, not just locomotives.

The *Roadmap* stated that in order to achieve its energy goals, the railroad industry would require $20 million per year in federal spending for 14 years. This, it was said, would bring public investment in railroad efficiency to the same level as for diesel truck efficiency. From economic and environmental perspectives, $280 million would have been a very small price to pay to revolutionize the way railroads use energy.

Regrettably, the federal funding was not forthcoming, and the railroads, locomotive manufacturers, and U.S. DOE never repeated the helpful, collaborative *Roadmap* exercise. The *Roadmap* was, however, followed by the DOE-sponsored "21st Century Locomotive Technology" program, which ran at least from 2003-2010. The stated goal of the 21st Century Locomotive Technology program was to "develop freight locomotives that (were) 25% more efficient by 2010 while meeting EPA Tier 2 emissions standards."[133] The program was also intended to increase railroad RTM/gallon index from 393 to 590 by 2020, a 50% improvement.[134] These efficiency goals, it turns out, were essentially the same as those articulated by the *Roadmap*.

The 21st Century Locomotive Technology program provided federal funding for GE locomotive R&D in these areas:

- Advanced diesel engine fuel injection with common rail technology (now in use in GE and EMD's Tier 4 freight locomotives)
- Advanced locomotive operating control system (which became GE's Trip Optimizer energy management system)
- Electric turbocharging
- A prototype hybrid locomotive using advanced batteries and energy management controls along with a *Hybrid* Trip Optimizer

DOE funding for locomotive efficiency improvement apparently ended at the conclusion of the 21st Century Locomotive

A June 12, 1978, *Railway Age* cover featuring a beautiful Don Ball, Jr., photograph of an Amtrak Turboliner speeding alongside the Hudson River on an Albany-to-NYC run. Amtrak operated a series of gas turbine-powered trainsets in the 1970s, 80s, and 90s. These attracted passengers but not fuel savings. The *Railroad and Locomotive Technology Roadmap* states that "The gas turbine has no potential for fuel savings compared with a diesel engine." Image Credit: *Railway Age*.

Technology program, and a summary report of accomplishments for the program could not be found. Nonetheless, federally supported research by diesel truck engine manufacturers like Cummins and Caterpillar has identified strategies to achieve clean diesel engines that are over 50% efficient—with projected efficiencies reaching as high as 60% using hybrid electrification and advanced waste heat recovery technologies.[135] Some truck engine efficiency technologies should be transferable to railroad locomotives, but there is no denying that the more demanding railroad operational environment limits some types of technology transfer. For example, narrower crankshaft bearings could improve prime mover efficiency by up to 1% but would not survive a coupler slack run-in or hard coupling. Different turbochargers and aftercooling arrangements might improve prime mover efficiency by more than 2%, but engines so equipped might be unable to make full horsepower at higher elevations or when operating in higher ambient temperature conditions including those encountered in long tunnels (which already can be problematic). Downsizing coolant pumps and other auxiliaries could reduce energy consumption but would adversely impact the longevity of engine components and the ability of locomotives to operate effectively under all conditions.

While improving the efficiency of diesel-electric locomotives may be more difficult than improving the efficiency of trucks, the railroads should take heed. The trucking industry is their main competitor, and in the coming years the fuel economy of commercial diesel trucks could increase from 7-8 mpg to 12-13 mpg. The latter is the fuel economy demonstrated by the U.S. Department of Energy's SuperTruck program.[136] Moreover, companies like Tesla are getting into the act. Automated electric trucks have the potential to revolutionize long-distance freight hauling (see sidebar in Section 4.2). To remain competitive, the railroads will need much more energy efficient locomotives and other systemic energy improvements.

Cresson, PA, June 2014. Photo by author.

CHAPTER FOUR

Energy Efficient Diesel-Electric Locomotive Operation

The *Railroad and Locomotive Technology Roadmap* observes that locomotive fuel use along the same route can differ by 12% to 20% among different crews equally familiar with that route.[137] This variance is reminiscent of a statement on a placard in the Railroaders' Museum in Altoona, PA, which quotes Ray Nycum, Pennsylvania Railroad's Road Foreman of Engines, speaking many years ago about the relative energy performance of steam locomotive firemen.

> **FIREMAN**
>
> *"One man would burn 40 ton of coal between Altoona and Harrisburg and another man wouldn't burn 20 with the same tonnage train."*
>
> – Ray Nycum, Road Foreman of Engines

Of course, much has changed since Mr. Nycum made that statement, including dieselization and the evolution of diesel-electric locomotives to their current advanced state. But a constant over the years is the role of skillful locomotive operation in achieving fuel economy.

4.1 Engineer Recommendations for Efficient Locomotive Operation

Here are general strategies that locomotive engineers (and increasingly, energy management systems) can use to improve fuel economy when operating diesel-electric locomotives:

- Operate locomotive in most efficient range as much as possible, i.e., the higher throttle settings (i.e., Notches 7 and 8).
- Pull trains at steady slower speeds when possible—a steady 15 mph may produce the highest ton-miles per gallon efficiency.[138]
- Don't start until the train automatic brake has been fully released.
- Accelerate slowly, advancing one notch at a time, as load stabilizes.
- Use throttle reductions and coasting to a stop, instead of maintaining speed and then stopping abruptly.
- Conserve train momentum and avoid stop-and-go operation whenever possible.
- Operate at a low throttle setting when traveling through rolling hills, allowing the descent down one hill to push a long train up the next hill.
- Stop on top of hill instead of at the bottom if there is a choice.
- Use throttle modulation to avoid bunching instead of *stretch-braking* or *power-braking*, a practice which involves actuating the automatic brake while a forward throttle setting is engaged

---

Train Speed vs. Fuel Consumption

"For any given train, the fuel consumption is related to its speed. Therefore, a reduction in speed will reduce fuel consumption. Conversely, an increase in speed will increase consumption."

~Fuel Conservation in Train Operation, Association of American Railroads, 1981

- Discontinue and return helpers as soon as possible to allow the locomotives remaining with the train to operate in fuel-efficient higher power notches.
- Take the following steps when operating in MU (multiple unit) mode:
  - Operate most efficient locomotive first.
  - Operate only enough locomotives to permit majority to operate in higher throttle settings.
  - Place assisting locomotives in idle or *isolate* them (turn them off) when their horsepower or dynamic braking assistance is not needed.[139]
- Minimize idling by following company idling policy, e.g., shut down locomotives when temperature is above 35 or 40 degrees, or use on-board anti-idling equipment.
- Use energy management systems like Trip Optimizer and LEADER when advantageous—consistent with safe operation of the locomotive and train.
- Report energy-consumption-related problems, as well as emissions-related problems such as smoke at the exhaust stack, to locomotive maintenance supervisors so that locomotives can be maintained properly to operate with peak efficiency.

As a matter of basic physics, Paul E. Rhine in his *Fuel-Saving Techniques for Railroads* recommended the steady 15 mph strategy to maximize fuel savings.[140] Slower speeds save fuel by keeping mechanical and aerodynamic friction to a minimum while also reducing rail and track maintenance costs.[141] Rhine confirmed this freight railroad efficiency strategy by conducting controlled main-line tests during his employment at Union Pacific, where he rose to Vice President-Operations as Manager of Train Energy Conservation. This 15 mph strategy should be employed while operating locomotives in their most efficient power ranges, i.e., high notches, and is listed here because, technically speaking, it

may be most efficient. Moreover, as the following paragraphs make clear, this rule has exceptions. For example, when going downhill the engineer should not apply brakes and needlessly sacrifice train momentum to reduce speed—unless otherwise instructed to do so.

Speeds in the 15 to 25 mph range have been known to cause "harmonic rocking" wherein freight cars with high centers of gravity rock back and forth, increasing wheel and rail wear, train drag, and chances of derailment.[142] This is not to say that operating cost effectively at slower speeds is impossible, but rather that care must be taken in order to avoid or minimize the possible pitfalls.

The general freight railroad rule that slower speeds save energy applies even when the speed comparison is in the higher ranges. Class I railroads have initiated fuel conservation programs that imposed train speed limits of 50-55 mph—though this restriction could be suspended when rail network needs temporarily required faster speeds.[143]

Moreover, it has been reported that years ago Conrail conducted an experiment that involved reducing the maximum speed of its intermodal trains between Chicago and New Jersey to 60 mph from 70 mph. The railroad found that this measure produced fuel savings of $4 million a year while only

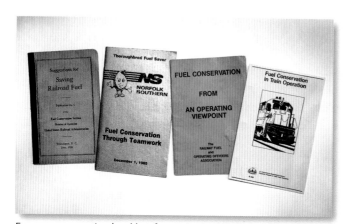

Energy conservation booklets for engineers. While the energy crises of the 1970s and energy consciousness of the 1980s motivated railroads to promote energy efficient locomotive operation, the training process began decades earlier during the steam era, as evidenced by the booklet *on the left* dated 1918 – all from the author's collection. Photo Credit: James Ulrich.

increasing average transit times by 1 hour.[144] With better traffic management and steadier speeds on that route, those savings might have been achieved without any loss of time.

But while slower speeds generally save energy, they can make bad situations worse on busy routes already congested with train traffic—thus reducing track capacity, an important concern to Class I railroads on key routes. When slower trains block faster ones, system velocity and efficiency is reduced. This adversely affects customer service and revenue. Slower system velocities also increase the cost of railcar hire and the time it takes cars to reach exchange points, as well as increase the number of locomotives and crew time required to move a given amount of product, as shown by the graph below. However, the converse may also be true. Depending on network conditions, slower train speeds can increase system velocity while significantly decreasing fuel use by avoiding stop-and-go operation.

A study Norfolk Southern undertook with GE around 2007 on trains operating between Chattanooga, Tennessee, and Macon, Georgia, clearly demonstrated that stop-and-go operation should be avoided whenever possible. Fuel consumption was calculated based on event recorder data and known notch fuel consumption of various locomotive models. The study found that when trains had to stop for meets in single-track territory versus running continuously toward their destinations, fuel consumption actually doubled.[145] This finding also supports a fuel economy argument for double-track operation.

Finally, it is worth pointing out that the Notch 7 and 8 throttle setting recommendation is not universally accepted. Prior to the widespread use of locomotive energy management programs, Union Pacific recommended (or mandated as a fuel conservation rule) that their engineers operate in throttle position 5 or below after their trains reached track speed.[146] The logic behind this recommendation was simple—less fuel is consumed in Notch 5 than in higher notches. Union Pacific achieved documented fuel savings with this approach even though higher notches produce more work per unit of fuel consumed.

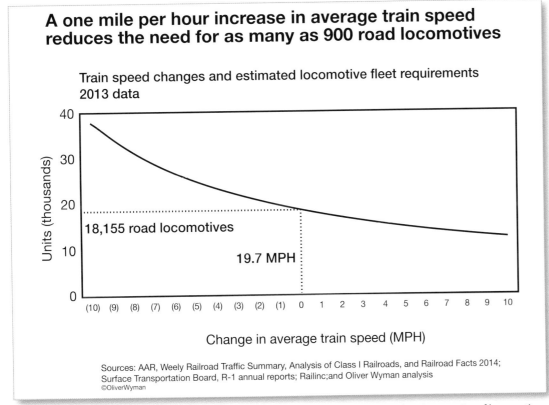

A one mile per hour increase in average train speed reduces the need for as many as 900 road locomotives

Train speed changes and estimated locomotive fleet requirements 2013 data

18,155 road locomotives

19.7 MPH

Change in average train speed (MPH)

Sources: AAR, Weely Railroad Traffic Summary, Analysis of Class I Railroads, and Railroad Facts 2014; Surface Transportation Board, R-1 annual reports; Railinc;and Oliver Wyman analysis
©OliverWyman

While potentially increasing fuel consumption, railroads can theoretically reduce their inventories of locomotives by operating trains faster. The relationship is not linear because faster trains require more horsepower per trailing ton. Image Credit: Oliver Wyman Group.

Incentive Programs for Locomotive Engineers

While some engineers take pride in operating their locomotives as fuel efficiently as possible, not all do. Moreover, engineers are motivated to move trains quickly across districts, yet efforts to do so can lead to energy inefficient operation. No wonder, then, that at various times railroads have created incentive programs to further engage engineers in the energy efficient operation of their locomotives.

A decade ago, Union Pacific and BNSF Railway—the Class I railroads that consume the most diesel fuel annually—implemented engineer incentive programs that they credited with significant fuel savings. In April 2006, UP announced that its "Fuel Masters" incentive program had produced a 5% reduction in fuel use in districts where it had been implemented in 2005.[147] These savings were estimated to be 16 million gallons of diesel fuel and $30 million—with expected savings to double in 2006. BNSF's "Fuel MVP" (Most Valuable Player) program was said to promote fuel conservation by encouraging a "friendly competition" among engineers. Wireless downloads (ERAD) were used to gather engineer performance data for that program.[148]

UP Fuel Master's program baseball cap and patch – from the author's collection. Photo Credits: James Ulrich.

FUEL CONSERVATION

UP Fuel Master energy conservation booklet for engineers – from the author's collection.

Generally speaking, incentive programs are a good idea but may be difficult to set up or make really effective. Here are some suggestions:

1. Incentive programs should not be deterred by the difficulty of determining objectively which engineers are doing the best job. There are many variables to take into consideration when evaluating energy efficient operation of locomotives and the performance of engineers. Train type, tonnage, routes, schedules, congestion, weather, etc., vary from shift to shift and can affect efficient operation. When Union Pacific set up its Fuel Master program, it decided to compare the locomotive energy use of participating engineers operating in the same territory—with the top 15% to 20% in each district receiving incentives. BNSF took a different tack with its Fuel MVP program. It compared the energy consumption of its engineers with their own past performances. Even though data gathering and analysis could be automated and done by computer, the analysis would always be somewhat imprecise because of the difficulty in fairly weighing all variables. Nonetheless, both programs worked.

2. Be inclusive and generous. When in doubt, incentive programs should reward a larger pool of engineers. That way, no one doing a good job would feel slighted. The incentive amount is also important. BNSF and UP both rewarded their fuel-efficient engineers with fuel debit cards that they could use to buy fuel for their own personal vehicles. BNSF reported that it distributed 3,100 personal fuel cards to its fuel-efficient engineers in 2008.[149] These $100 and $50 fuel cards were minimal incentives. If they worked well, they did so because they provided a token of appreciation from the company that the engineers valued. Union Pacific incentives were reported to be more generous in order to get greater results. In the end, engineer enthusiasm for fuel economy also depends on good will, altruism, and environmental concern as well as continued positive reinforcement for individual engineers.

3. Incentive programs work best when reinforced by engineer training and a comprehensive energy awareness program. This was recognized by UP, which multiplied the effectiveness of its Fuel Master program with accompanying posters, flyers, and brochures, educating engineers how to operate more efficiently. Peer trainers were used. A Fuel Masters Simulator Challenge, which invited engineers to compete on simulators, was added to the real in-cab competition for incentive rewards.

4. Incentives can be undermined by unaddressed disincentives. For example, while operating at a steady slower speed generally saves energy and generally should be rewarded, what if such operation extends an engineer's workday? Or prompts management warnings about trains moving too slow or causing delays? Disincentives like these exist. With them in place, most engineers will opt for faster speeds and greater fuel use.

5. Incentive programs depend on good labor relations. A cooperative spirit between railroad managers and engineers is needed to make incentive programs work. BNSF recognized this and signed an agreement with the Brotherhood of Locomotive Engineers and Trainmen on December 3, 2007, to gain the union's support for its Fuel MVP program.

Both BNSF and UP programs have wound down and ended, hopefully with conservation lessons learned and institutionalized. Energy management software now installed in locomotives played a part in ending these incentive programs because, to an increasing extent, these systems have taken over the energy efficient operation of locomotives. Nonetheless, the engineer still plays a key role. As long as that is the case, it makes sense to train, encourage, and reward engineers who operate their locomotives efficiently.

4.2 Role of Energy Management Systems

"Cruise Control" for Locomotives

GE's Trip Optimizer and New York Air Brakes' LEADER systems save energy by introducing a degree of automated locomotive operation. Both have sophisticated algorithms and advanced capabilities designed to save energy in a simple way—by conserving train momentum.[150] This can be accomplished in large measure by eliminating unnecessary braking, since braking turns mechanical energy into useless waste heat, which, in turn, must be replaced by diesel fuel to bring a train back up to speed. Thus, these programs use throttle and dynamic brake settings to keep trains going at modest speeds that can be maintained without slowing or stopping, and that will have minimal adverse impact on scheduled arrivals. While "cruise control" is commonly used to describe Trip Optimizer and LEADER, the term is misleading if it is taken to imply constant speed; these programs modulate throttle and dynamic brake settings,

changing speed en route. (Note that these energy management programs were also discussed in Section 2.3.)

Just how good are Trip Optimizer and LEADER? There is no doubt that Class I railroads believe they improve fuel economy. This is demonstrated by the near-total buy-in and increasing use by the railroads. But can they outperform the most skilled locomotive engineers? GE and NYAB do not make this claim, and the author has been told by Class I railroad locomotive managers that the best engineers can

Engineer J. F. Brunson in the hot seat of Atlantic Coast Line's train 75, the *Havana Special*, in August 1957. The locomotive engineer's job has changed considerably since the days when EMD E-unit locomotives were a regular sight on the rails. Photo Credit: William D. Middleton.

"beat the machine."[151] However, these managers also told him that the average engineer cannot operate a locomotive (or consist of locomotives) more efficiently than energy management programs, and that it's difficult for even the best engineers to beat these systems consistently, every day. While outperforming engineers on average, it's also said that the energy management programs reduce variance in engineer performance[151]—a management objective.

The ability of Trip Optimizer and LEADER to save energy and energy dollars may be undisputed, but they pose new challenges. While younger engineers may more easily adapt to monitoring these energy management systems rather than hands-on locomotive operation, at least some older, experienced engineers have pushed back, taking the position that Trip Optimizer and LEADER are locomotive automation systems that present safety concerns. These concerns led the Brotherhood of Locomotive Engineers and Trainmen (BLET), on February 4, 2016, to ask the Federal Railroad Administration to issue an emergency order banning the use of these energy management systems.[152] The union criticized Trip Optimizer and LEADER, arguing that these systems:

- Cannot anticipate all track conditions, or respond to emergency conditions

- Will suddenly turn off and return locomotive operation to the engineer when train operation requires immediate attention—even though the engineer may be distracted, bored, or lulled into "automation complacency"[153]
- Compete for the engineer's attention, which must be available for other tasks
- Are often mandated for use a certain percentage of time, e.g., 85%, which is not possible when "chasing signals," encountering work crews, etc.,—with instances of non-use requiring en route documentation to avoid disciplinary action
- Weaken engineers' operating skills and judgment by reducing practice in actual locomotive operation

The BLET complaint prompted a June 17, 2016, response from the FRA that promised to conduct research on all BLET concerns while allowing for the continued use of these energy management systems.[154] As of May 2018, there was no public progress report available on these matters.

Railroad management tends to view the BLET complaint as an expression of resistance to change, which of course it is in part, though the specific concerns raised by BLET have validity and deserve consideration. While computers fly airliners successfully, pilots must be ready and able to take control when an error occurs. Yet pilots do lose attentiveness as well as flying skills when merely monitoring computer screens.[155]

An unspoken element in this debate is crew size. The railroads argue that a one-person crew—the engineer—can safely handle increasingly automated locomotive and train operation, while railroad workers unions, with support from the FRA during the Obama Administration, maintain two-person crews are still needed.

While full automation of main-line locomotive operation has not yet been achieved in the U.S., the industry is moving in this direction. Once PTC is coupled with advanced energy management systems, safer and more energy efficient operation may be possible with only a single operator in the cab, and, in all likelihood, the operator will be handling the controls less.

Automation technology for certain types of railroad operation already exists. Nearly 50 city passenger-rail systems and many smaller people-mover systems are automated.[156] The Australian railroad Rio Tinto has begun operating automated "engineer-less" long-haul freight operations in the Australian outback, and announced that by the end of the first quarter of 2018 65% of its *train kilometers* were being operated in autonomous mode.[157] In 2016, Deutsche Bahn, citing shortage of engineers, announced that it envisioned running automated trains within 5 years.[158] Then, in 2018, began an early trial of automated train operation (with driver aboard to supervise) along a 100 km purpose-built route in the Netherlands using Alstom technology.[159] Given the likelihood of eventual competition from driverless trucks, some level of automated freight locomotive operation appears to be inevitable[160]—though it remains to be seen how railroads will effectively function with reduced operating staff as long as trains are subject to mishaps like broken couplers and air hoses that lend themselves to repair by two-person crews.

A March 29, 2018, FRA solicitation for input on U.S. railroad automation[161] produced numerous critical comments. Arguments against full autonomous train operation included the ability of two-person crews to respond to emergencies and operational problems, the importance of railroad jobs, and the risk of automated locomotives being hacked[162]—a concern expressed shortly after press accounts revealed that Russian hackers had gained access to U.S. infrastructure including nuclear power plants operation.

Advances in Semi-Truck Technology Challenge Railroads

On November 16, 2017, Tesla chief executive, Elon Musk, threw down the gauntlet when he unveiled the Tesla super truck, which he claimed would begin production by late 2019 and revolutionize the trucking industry. Significantly, Musk also claimed that his super truck would surpass the economics of freight railroads for long-haul shipping.[163]

According to Musk, the Tesla semi-truck would be:

- Electric—Powered by electric motors and batteries
- Self-driving—Operated by computer with a "driver" on-board to monitor systems
- Energy efficient—Recovers kinetic energy with regenerative braking
- Solar-powered—To be recharged by a national network of quick-recharging stations fed by solar power installations

- Long range—Capable of traveling 500 miles on a single battery charge
- Fast—Able to accelerate to 60 mph in 5 seconds without a trailer and in 20 seconds when carrying a load of 80,000 pounds
- Safe—"Ten times safer" than a truck driven by a human
- Clean—Zero emissions
- Easily maintained—No engine to maintain; little conventional brake wear
- Convoy-capable—Highway operation in "trains" of three self-driving trucks
- Inexpensive—To be 17% less costly to own and operate than diesel trucks, and 42% less costly when operating in three-truck convoys
- Guaranteed—Drive train guaranteed for 1 million miles

This impressive list of features demonstrates Musk's visionary nature. It remains to be seen whether he and his company can deliver on these promises—though in other areas Tesla has produced astonishing results. Affordable batteries with sufficient energy and power density for heavy trucking must still be commercialized.

However, transferring freight from rail to road using large, fast-moving self-driving trucks will meet substantial public resistance,[164] at least in the short term. Consider:

- The safety of self-driving trucks is unproven. It has not been demonstrated that sensors and software can match the skills of a skilled truck driver under all conditions.

Tesla automated electric super truck slated for production in late 2019. Photo Credit: Tesla.

- Well-publicized accidents, implicating autonomous vehicles, are inevitable and will undermine public confidence.
- The dangers associated with possible computer failure or the hacking of self-driving truck control systems have yet to be addressed.
- Putting large numbers of these trucks on public highways will exacerbate traffic problems, adversely impacting automobile use and conventional commercial traffic.
- Drivers of conventional vehicles, especially smaller passenger vehicles, are unlikely to feel comfortable sharing the highway with large self-driving trucks, individually or in convoys.
- Greater volume of heavy truck traffic will increase road repair and maintenance costs.
- Highway expansion required to accommodate a proliferation of these trucks would further shift costs to the general public.
- Autonomous trucks threaten many thousands of truck driving jobs.

Despite challenges, the eventual production and use of these vehicles in some form and on some routes appears to be likely within the next 10 years. This type of automated, potentially environmentally responsible trucking technology—whether introduced by Tesla or other companies—will challenge the railroad industry to "innovate or retrench."[165] What is needed is a competitive response, especially for intermodal traffic. Quick change, however, has traditionally been difficult for the railroads, in part because of the industry's conservative management style and the long life of motive power assets—its diesel-electric locomotives.

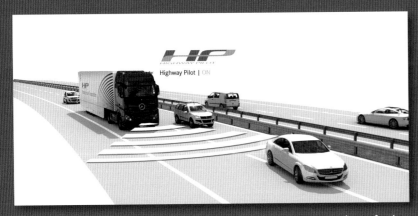

Highway Pilot and Mercedes-Benz Future Truck 2025. A number of other truck manufacturers are developing and testing autonomous trucks, including Mercedes and Freightliner, both divisions of Daimler Trucks. Image Credit: Daimler AG.

Locomotive Black Box

The locomotive event recorder is like an airplane's black box; it records control settings and performance parameters when a locomotive is in use, permitting this information to be reviewed afterwards. Event recorder information can be provided to railroad system managers while trains are en route via cellular communication or Event Recorder Automatic Download (ERAD) systems that send data to managers whenever a train passes wayside relay stations,[166] also called *wireless access points*. (Note that event recorders were also discussed in Section 2.3.)

Needless to say, event recorder information is of special interest if there has been a problem or accident, though event recorders have also been used by railroads to assess engineer fuel economy performance. This is possible because these devices gather data on aggregate throttle notch times and braking. This information can show managers whether engineers are operating in the most efficient notches. The event recorder also shows if an engineer engages in practices like stretch braking, which—as explained—wastes fuel and prematurely wears brake pads on all cars in the train. Significantly, event recorders show to what extent the engineer has Trip Optimizer or LEADER engaged, or (if operating in "prompt" mode) how often the engineer is following the prompts.

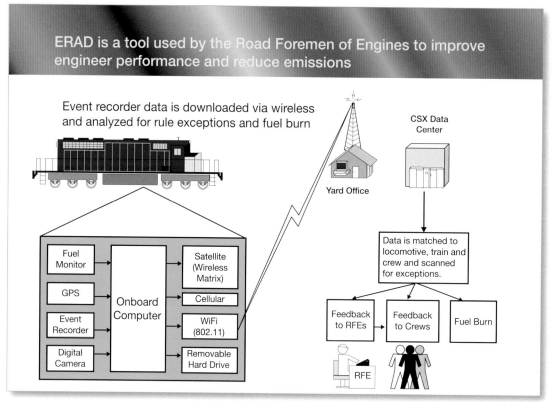

Here's how ERAD could work as an energy conservation training tool. Image Credit: CSX Corporation.

Thus, ERAD systems provide feedback on engineer performance that can be used to encourage and instruct engineers to operate their locomotives and trains more safely and efficiently. This objective is not achieved, however, if engineers view this technology negatively as a "got-cha" device for gathering information to be used against them, and railroads use it punitively.[167] A positive relationship between management and operating staff could ameliorate these concerns.

Video recorders now accompany event recorders in locomotive cabs. These special DVRs collect data from outward- and inward-facing video cameras.[168] Depending on how they are programmed, these devices may begin recording when the locomotive starts moving and can also record audio inside the cab, outside it (to document the use of the locomotive horn), or both. The railroad industry's interest in gathering this data to ensure that engineers are paying full attention while operating locomotives is understandable, but cooperation on energy conservation and other matters could decline if surveillance is constant and resented.

Like it or not, the engineer profession is undergoing a digital revolution. It may be harder for engineers to take pride in their profession when computers operate "their" trains and management can monitor all their actions and behavior. Some adjustment and accommodation on both sides is required. For example, railroads could allow their best locomotive engineers more opportunity to fully operate their locomotives, or to operate energy management systems in their "driver-prompt" or "driver-assist" modes. A full digital transition would be a mistake.

4.3 Dispatcher and Trainmaster Recommendations

The locomotive engineer and on-board energy management systems can only do so much to save energy. Energy efficient operation requires the full participation of a team of railroad professionals who are responsible for assigning locomotives, building trains, and operating them across a region. Their contribution to locomotive energy conservation can be as great as the engineers. Here are some energy efficient strategies for the rest of the operations team to employ:

- Select the most energy efficient locomotives for operational service
- Assign the most efficient locomotive to be lead locomotive for any given train
- Assign the correct amount of total horsepower to trains, so that peak loads can be met with locomotives operating in their most energy efficient throttle settings (Notches 7 and 8)
- Schedule and route trains to avoid congestion that results in energy-wasting stop-and-go, catch-up operation plus additional idling
- Schedule trains to avoid "stop and proceed" signaling or train orders – though this type of operation may be unavoidable in districts with single track and passing sidings
- Schedule train arrivals to avoid yard traffic congestion
- Schedule trains to operate at steady modest speeds in order to minimize stop-and-go operation and higher levels of mechanical friction and aerodynamic drag
- Use meet planner and the network efficiency software to accomplish the above scheduling tasks
- Warn engineers of upcoming congestion or speed restrictions so speed can be reduced by coasting
- Use Centralized Traffic Control (CTC), Computer Aided Dispatching (CAD) and Positive Train Control (PTC) systems to accomplish the above

- Shorten yard dwell times to permit slower operating speeds on the road without incurring travel time penalties
- Avoid return trips with empty cars when possible
- Build trains to minimize air friction, i.e., if possible, arrange trains with cars of equal heights (for intermodal trains, ensure full "slot utilization" where double-stacked cars are maximized)
- Implement the use of add-on aerodynamic devices like Union Pacific's ArroWedge (called an *aerodynamic pseudocontainer*[169] in patent filing) if available and practical[170]

4.4 Railroad Management Recommendations

Here are a few strategies for management to implement to encourage more efficient locomotive operation:

- Make a top-level corporate commitment to energy conservation and environmental sustainability
- Maintain an energy efficient locomotive fleet
- Train, support, reward, and maintain a positive relationship with engineers and operating staff to facilitate good will and energy efficient operation

- Hire at least one full-time fuel economy or energy conservation officer who reports to top management (near VP level) and, with the help of support staff, promotes company-wide energy-saving
- Maintain a positive relationship with operating divisions, regions, and employee unions to plan and secure support for energy conservation efforts
- Respond constructively to engineer concerns about Trip Optimizer and LEADER
- Recognize that reducing the company operating ratio (costs/revenues) by cutting fuel economy and energy conservation programs and staffing is counter-productive

4.5 Distributed Power Operation

Imagine trains nearly 3 miles long, with 250 cars, weighing as much as 20,000 tons.[171] While trains like these may still be the exception, the railroad industry is increasingly consolidating trains, running them longer and heavier.

Distributed power (DP) is a mode of operation defined by using multiple groups of locomotives in a train. In addition to locomotives pulling from the front of the

Railroads use simulators to train engineers in the safe and efficient operation of locomotives. This one, manufactured by CORYS, is used by CSX. Photo credit: CORYS.

train, DP trains will have rear pushers, and possibly locomotives located in the middle of the train, which push the cars in front of them and pull the ones behind them. When a single unit or consist of remote locomotives is used in a long heavy train, the ideal positioning of the remote-control units is generally two-thirds back in the train from the head-end—thus, equalizing the distribution of in-train forces.

Long heavy trains need DP to operate safely and efficiently because trailing tonnage could otherwise create drawbar forces in draft (slack) conditions that exceed coupler strength. These forces, which increase when trains accelerate or are pulled up gradients, can break a train in two. Without DP, long heavy trains can also encounter buff (compressive/bunching) forces extreme enough to knock couplers out of coupler pockets. These conditions may be encountered when braking, descending grades, or traversing undulating routes.

During the steam era, rear pushers were common, though mid-train locomotives were not. But unlike steam locomotives, which required a full crew for each locomotive, multiple diesel-electric locomotives operating in DP mode require just one two-person crew in the lead locomotive.

Distributed power makes longer trains possible with existing coupler strength but requires radio connectivity, which is

DISTRIBUTED POWER

Conventional Train: all locomotives in "head end"

DP with additional locomotives at rear (pushers)

DP with additional locomotives in middle and rear

Image Credit: Robert Hochberg.

occasionally a problem in deep canyons or when trains pass through long tunnels. This connectivity is supplied by telemetry systems like GE's Locotrol, which was originally developed in the 1960s and now can control up to four locomotive groupings or consists located throughout a train. These consists may be operated in unison or in independent mode, where the latter mode allows a long train cresting a hill to ease up on motive power in the front while continuing to apply full power in the rear. This improves train handling and energy efficiency. LEADER's independent dynamic brake mode is called *Asynchronous Distributed Power*.[172]

Distributed power makes better use of dynamic braking by spreading out braking forces to the front, middle, and rear of the train. Distributed dynamic braking is a real asset when descending long,

curving gradients on mountainous routes. Automatic train brakes also benefit from DP because brake pipe pressure can be reduced or increased simultaneously in more than one location within the train. This has the effect of reducing brake application and release times. It also speeds up railcar air reservoir recharging.

The use of higher tractive effort AC locomotives with DP has made it possible for fewer locomotives to move longer trains over a variety of terrains. Distributed power also saves energy by increasing train controllability and reducing wheel/rail friction when the train is negotiating curves. There is less wheel grinding and squealing, because when mid-train engines push the cars ahead of them, it relieves some of the lateral force or sideways pressure pushing car wheel flanges against the rails.

Longest, Heaviest Trains

In 1967, the Norfolk & Western Railroad (now part of Norfolk Southern) decided to push the limits of what would later be called distributed power, and set a world's record for the longest, heaviest train.

After a number of record-setting runs in October and November of that year, N&W ran the train that still holds the U.S. record. Dubbed the "Super Train," it consisted of 500 loaded coal cars, weighing 48,170 tons and stretching 4 miles long. It ran on the slightly downhill route of 47 miles between Iaeger, WV (elevation 983 feet), and Portsmouth, OH (elevation 533 feet). Motive power was 21,600 horsepower—three 3,600 hp EMD SD45 locomotives on

lead and three remotely controlled SD45s at the 300-car mark.[173] When starting, the lead locomotives moved ahead for 2 minutes and 12 car lengths before the rear of the train moved!

The current world record for the longest train is held by BHP Billiton Limited. On June 21, 2001, between Newman and Port Headland in Western Australia, this Australian mining company ran a 682-car train that weighed 99,734 tons and was 4.57 miles long. Its cargo was iron ore. Motive power of 48,000 hp was supplied by eight 6,000 hp GE AC6000 locomotives placed in five locations in the train.[174]

N&W's 450-car record-breaking train winds through the Appalachian Mountains on the cover of Norfolk and Western Magazine, November 20, 1967. Image Credit: © Norfolk Southern Corp.

The story of the final record-breaking 500-car "Super Train" is detailed in the January 1, 1968, edition of same company magazine. Image Credit: © Norfolk Southern Corp.

Union Pacific, known for pulling heavy freight trains for long distances, reports that 70% of its ton-miles of freight are now hauled with DP trains. While published reports demonstrating the fuel economy benefits of DP are not available, UP reports that distributed power reduces fuel consumption by 2.5% to 5%.

Longer trains require longer sidings, and building trains with mid-power DP can present considerable challenges to yard crews—primarily limiting the practice to certain territories. The additional investment of terminal switching and physical plant resources may pay for Union Pacific or BNSF—because they have more double track (making the need for longer sidings less an issue) and many of their trains run intact very long distances cross-country—but not for Eastern railroads. Longer trains can also pose safety concerns to the public because they block crossings for longer periods of time, potentially delaying emergency response vehicles or providing greater temptation by motorists and pedestrians to disobey and circumvent crossing gates.

Norfolk Southern uses DP locomotives located at the front and end of trains, often assisted by additional front and rear end "helpers" (whose consists are operated by additional engineers). These additional locomotives pull and push heavy freight trains over the Pennsylvania Allegheny Mountains and also

Rear-end diesels pushing mile-long oil train up to Gallitzin Tunnel through Cassandra, PA, 2014. Photo by author.

assist with dynamic braking when going downhill. Helpers make sense when many trains are going over a relative short section of railroad where steep grades are concentrated. Adding remote-control DP locomotives to all trains on that route would unnecessarily consume a large amount of locomotive asset because the DP locomotives would only be used for a short distance.

In a departure from past practice, CSX announced in 2017 that it will operate

in mountainous areas without helpers.[175] However, this policy decision may still be under review.

Distributed Power Energy Efficiency Opportunities. It might be possible to further reduce wheel/rail friction on curves by placing additional locomotives throughout the train—though potential savings could be eroded if this strategy resulted in locomotives operating at lower power outputs, and therefore outside of their optimal efficiency range. This strategy also would

119

involve more time assembling trains, reducing its practicality.

On straight, level track when additional horsepower is unneeded, as many MU-ed DP locomotives as possible should be left in idle or turned off so the remaining locomotives can operate their prime movers at high throttle settings with greater efficiency.

DP locomotives operating in mountainous areas could be configured with large battery systems or battery tenders so they could capture and store for later use the electricity generated by dynamic braking. (See battery tender discussion in Section 6.3.)

Norfolk Southern pushers fly back to Altoona for another assignment, February 2017. Photo by author.

Tier 4 leading on Tehachapi Pass. Photo by Mike Danneman.

122

CHAPTER FIVE

EPA Emissions Standards, TIER 4 Locomotives, and Energy Impact

The U.S. Environmental Protection Agency (EPA) has been administering exhaust emissions standards for vehicles with internal combustion engines, including diesel-electric locomotives, for over 20 years.[176] The rules are complicated, but they have been effective in reducing air pollution. As we shall see, however, clean air comes at an energy price.

5.1 EPA Locomotive Emissions Regulations Overview

Diesel-electric locomotive emissions standards follow this general outline:

- There are five "tiers" or series of regulations, Tier 0 – Tier 4.
- Each successive tier applied more significant emissions reduction requirements to new locomotives, though the initial rulemaking applied to locomotives manufactured from 1973.
- Initial regulations for Tiers 0 – 2 became effective in 2000 and 2001.
- New regulations issued in 2008 reduced emissions for Tiers 0 – 2 effec-

tive in 2010 (thereafter referred to as Tier 0+, Tier 1+, and Tier 2+) and established Tier 3 and 4, which became effective in 2011/2012 and 2015.
- When older locomotives have their engines overhauled, they must be upgraded to meet more stringent standards.
- Regulated emissions are nitrogen oxides (NOx), particulate matter (PM), hydrocarbons (HC), carbon monoxide (CO), and smoke.
- These emissions contribute to air pollution and respiratory illness, and CO is a poisonous gas.
- Emissions of NOx, PM, HC, and CO are measured in grams per BHP-hr (amount of pollution per unit of engine work) while smoke is measured according to an opacity test.
- Emissions are regulated based on an EPA-established "duty cycle," i.e., a pattern of throttle settings determined to be representative of actual average operation.
- Line-haul and switcher locomotives have different duty cycles and emissions standards.
- New locomotives must initially perform better than their tier's

emissions standard so that they will meet the standard when retested later in their operational lives.
- The EPA's Averaging, Banking, and Trading (ABT) program permits manufacturers to sell new *credit locomotives* or *credit-using locomotives* that don't meet the current emissions standards if the additional emissions these locomotives would produce are offset by "banked" certified emissions reductions from earlier locomotives that exceeded the EPA's requirements at the time they were manufactured. It's also possible for certified emissions reductions to be purchased, or to be traded with another manufacturer.[177]
- A Tier 5 has been proposed by California.
- While the EPA has the legal authority to regulate carbon dioxide as an air pollutant, it has not yet included this greenhouse gas in locomotive emissions standards.

As indicated above, the most recent and rigorous locomotive emissions regulations are the Tier 4 standards that went into effect January 1, 2015. These standards were substantially more stringent and challenging than for previous tiers. For example, Tier 4 locomotives are required to emit 76% less NOx and 70% less PM than Tier 3 locomotives.[178]

Locomotive and diesel engine manufacturers have used different strategies to meet Tier 4 emissions standards: Some locomotive models have achieved that reduction without aftertreatment while other models have used aftertreatment.[179] U.S. Tier 4 locomotive builders are listed here, with their engine manufacturer shown in parentheses:

- **Without Aftertreatment (freight locomotives)**
  GE (GE)
  EMD (Caterpillar)

- **With Aftertreatment (passenger locomotives or genset switchers)**
  EMD (Caterpillar)
  Siemens (Cummins)
  MotivePower (Cummins)
  Brookville (Cummins)
  National Railway Equipment (Cummins)
  Railpower (Cummins)
  Railserve (Cummins)
  Tractive Power (Caterpillar)
  Knoxville Locomotive Works (MTU)

The cost of a new Tier 4 freight locomotive without aftertreatment is reported to be approximately $3.5 million[180] while the cost of a Tier 4 passenger locomotive with aftertreatment is said to be $6.5-$7 million.[181] The cost differential between Tier 4 freight and Tier 4 passenger locomotives is not attributable to the cost of aftertreatment. Passenger locomotives always cost more, in part because they (a) are required to meet more rigorous crashworthiness standards, (b) operate at higher speeds, (c) provide head-end power to the passenger cars they pull, and (d) are often subject to expensive consultant-driven customized specifications. The California Air Resources Board estimates the cost of the aftertreatment system alone to be $250,000.[182]

5.2 Achieving Tier 4 Compliance Without Aftertreatment

The EPA recognized that Tier 4 would be technologically challenging for locomotive manufacturers and assumed that they would use selective catalytic reduction (SCR) aftertreatment technology to meet the new emissions thresholds. However, U.S. Class I freight railroads rejected the aftertreatment option because of the potential space requirements of the aftertreatment technology, the logistics and infrastructure involved in handling and supplying the urea chemical reductant used by the aftertreatment process (see aftertreatment discussion in Section 5.3), and its cost.[183] The price tag of a nationwide urea infrastructure was estimated to be $1.5 billion.[184] Accordingly, the railroads put pressure on GE and EMD to find another way to build Tier 4-compliant road diesel-electric locomotives. GE was able to accomplish that by the 2015 deadline with an upgraded ES44AC locomotive, its ET44AC. In a major setback, Progress Rail's EMD found that it could not meet Tier 4 emissions levels with its existing two-stroke engine without incurring large fuel economy losses.[185] Accordingly, the introduction of its Tier 4 locomotive was delayed for 2 years, while the company developed a new four-stroke engine, with locomotive production beginning the fourth quarter of 2016.[186] Accommodating the weight of this new engine was reported to be a challenge.[187]

General Electric established a goal of producing a Tier 4-compliant locomotive that was also 3% more energy efficient. Achieving this goal would be challenging, given that both GE and EMD were concerned that compliance would *decrease* energy efficiency. GE has not announced whether it attained this energy goal for the ET44AC, and it apparently has not (see Section 5.4). GE was successful, however, in meeting Tier 4 emissions standards without the use of any type of aftertreatment.[188]

GE and EMD Tier 4 locomotive print advertisements, *Railway Age*, April 2016. Image Credits: GE Transportation and Progress Rail, A Caterpillar Company.

The ET44AC uses a variety of advanced technologies to operate efficiently while meeting Tier 4 emissions standards:

- A high-pressure common rail fuel delivery system
- A higher engine compression ratio and two-stage turbocharging
- Exhaust Gas Recirculation (EGR)
- Cooling system capacity increase of 25% yielding 50% more heat rejection
- Improved engine computer controls with 50% more sensors

- Elimination of the secondary (auxiliary) alternator
- Inverter controlled auxiliaries

Common rail fuel delivery systems optimize combustion. They obtain their name from pipe manifolds (common rails) on each side of the engine that supply highly pressurized diesel fuel to individual cylinder fuel injectors. The fuel is pressurized by a special high pressure fuel pump to as much as 35,000 psi. This pressure effectively forces the fuel into the injectors

at near-constant high pressure. There is no ramping up to high pressure as occurs with individual cam-driven injection pumps. High-pressure common rail also increases the atomization of diesel fuel when it is injected into engine cylinders. This improves the fuel's mixing with oxygen – resulting in more complete combustion, improved fuel economy, and reduced particulates emissions. Double-wall piping (a pipe within a pipe) may be used to protect mechanics and operating crews from the hazards of a manifold leak.

Increasing the compression ratio and employing compound-turbocharging improve power output and energy efficiency. Compound-turbocharging places multiple turbochargers in series—with engine exhaust first powering the high-pressure turbocharger before being directed to the low-pressure turbocharger. This arrangement progressively extracts more energy from the exhaust waste stream. Combustion air follows the reverse path, initially being compressed by the low-pressure turbocharger. Then it is cooled before being further compressed by the high-pressure turbocharger, after which it is cooled again before being thrust into the engine's cylinders.

Unmodified diesel engines produce excessive amounts of NOx because they burn with excess air (oxygen) at a high flame temperature. Exhaust Gas Recirculation (EGR), which

125

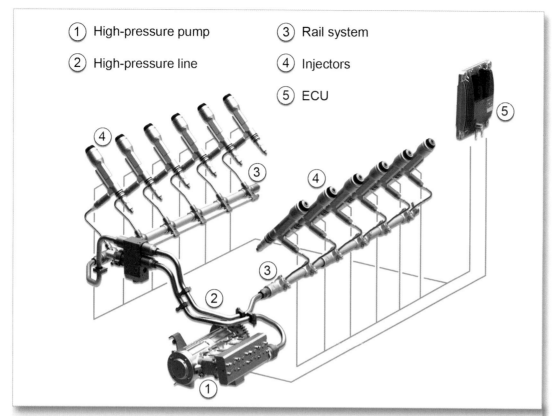

① High-pressure pump
② High-pressure line
③ Rail system
④ Injectors
⑤ ECU

High-pressure common rail fuel injection system. The *ECU* or Engine Control Unit shown in the diagram regulates the amount of fuel injected into the cylinders as well as injection timing and sequencing. Common rail fuel injection on MTU engines, for example, involves distinct multiple *fuel injection events* per power stroke: (1) the *pre-injection event* initiates combustion, (2) the *main injection event* produces engine power, and (3) the *post-injection event* minimizes particulates. MTU is the core business of Rolls-Royce Power Systems. Image Credit: MTU.

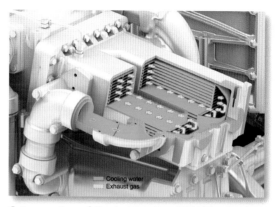

Cutaway view of the inside of an MTU Tier 4 prime mover EGR cooler. Exhaust gas drawn off for recirculation has a temperature of approximately 1,200°F—far too hot to be fed directly into the cylinders. By transferring much of this heat to engine cooling water, this EGR cooler reduces the temperature of these gases to around 250°F. The EGR cooler depicted here is designed to fit within the existing engine profile in order to minimize its space requirement. Image Credit: MTU

injects a portion of cooled exhaust gas back into the engine's cylinders, reduces NOx formation by tackling both conditions. It reduces excess oxygen in the cylinders and cools off combustion temperatures. But EGR operation adversely affects particulate emissions, horse-power, and energy efficiency. EGR's engine efficiency losses are significant, so making up those losses through other design features is a challenge.[189] Combining EGR with high-pressure common rail fuel injection compensates for EGR's drawbacks to some extent.

EGR requires additional cooling capacity to cool the recirculated exhaust gas prior to injecting it into the cylinders. This is accomplished by an EGR cooler, a gas-to-liquid heat exchanger, which extracts heat from the exhaust gases and transfers it to water circulating within the engine's cooling water system. That cooling system must be enlarged to handle the additional heat rejection. A Tier 4 locomotive equipped with EGR releases more engine waste heat through its radiators than exhaust stack. That is the opposite of previous locomotive types where stack waste heat exceeded radiator waste heat.[190]

EMD'S SD70ACe-T4 takes a similar approach with common rail fuel delivery,

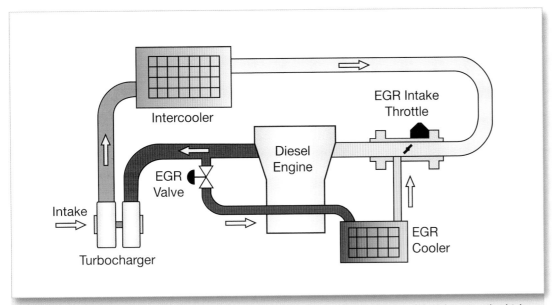

Simplified schematic showing combustion air and exhaust flows in a turbocharged diesel-powered vehicle with exhaust gas recirculation (EGR). The EGR valve controls the amount of exhaust gas that is recirculated. That gas, like the turbocharged combustion air, must be cooled prior to entering the engine's cylinders—hence the EGR cooler and the combustion air intercooler. Image Source: DieselNet.com.

5.3 Urea SCR Aftertreatment to Achieve Tier 4 Compliance

In contrast to the freight railroads, new passenger locomotives achieve Tier 4 NOx compliance with the use of urea selective catalytic reduction aftertreatment. These locomotives include the Siemens SC-44, EMD F125, and Motive Power MP54AC.

Prior to production of Tier 4 locomotives with aftertreatment, these challenges were identified[193] concerning the application of this technology:

- Identification of aftertreatment technology
- Space availability within the locomotive (called a *packaging* issue)
- Control, especially during transients, when load and horsepower change
- Impact of exhaust temperature (SCR needs hot exhaust to work effectively)
  - Cold starts
  - Low temperature operation
- Fuel economy loss
- Infrastructure requirements
- Maintenance and reliability issues
- Cost

Urea aftertreatment technology passes exhaust gases and ammonia (in the form of

compound-turbocharging (actually using three turbochargers, one for low engine speed and two for high speed), EGR, larger cooling system, and inverter-driven accessories. Like the GE ET44AC, the EMD SD70ACe-T4 uses a four-stroke diesel engine and individual inverters and controls of each traction motor and axle—major departures from EMD's past practice. Recognizing that a four-cycle engine would weigh more than its predecessor two-cycle, EMD looked for ways to reduce weight in other areas. Substituting fabricated trucks for those with a cast frame, for example, saved

1,000 pounds per truck.[191] To comply with Tier 4 particulates standards, the SD70ACe-T4 engine is reported to use particulate oxidation catalyst aftertreatment.[192]

Hot exhaust creates a very harsh environment. Not surprisingly, engines with EGR may need more maintenance because of soot accumulation on the valve that controls EGR gas flow, and on the heat exchanger surfaces in the gas-to-liquid heat exchanger (cooler) used to lower the temperature of the hot exhaust gases being recirculated.

127

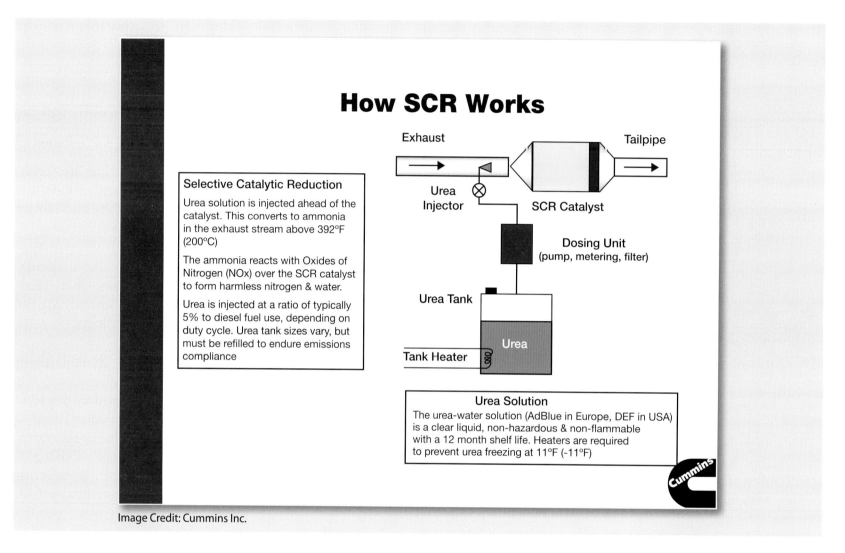

How SCR Works

Exhaust

Tailpipe

Urea Injector

SCR Catalyst

Selective Catalytic Reduction

Urea solution is injected ahead of the catalyst. This converts to ammonia in the exhaust stream above 392°F (200°C)

The ammonia reacts with Oxides of Nitrogen (NOx) over the SCR catalyst to form harmless nitrogen & water.

Urea is injected at a ratio of typically 5% to diesel fuel use, depending on duty cycle. Urea tank sizes vary, but must be refilled to endure emissions compliance

Dosing Unit
(pump, metering, filter)

Urea Tank

Urea

Tank Heater

Urea Solution
The urea-water solution (AdBlue in Europe, DEF in USA) is a clear liquid, non-hazardous & non-flammable with a 12 month shelf life. Heaters are required to prevent urea freezing at 11°F (-11°F)

Cummins

Image Credit: Cummins Inc.

urea-based Diesel Exhaust Fluid or DEF[194]) over a catalyst, which is a substance that stimulates a chemical reaction without using up itself. The reaction looks like this:

$$NOx + Ammonia + SCR\ Catalyst \rightarrow Nitrogen + Water$$

This process can reduce NOx in diesel exhaust gas by as much as 90% and thus is able to meet Tier 4 requirements. However, approximately 1 gallon of DEF is required for every 20 gallons of diesel fuel. At that rate, a large diesel-electric locomotive carrying 5,000 gallons of diesel fuel would need to also carry, and regularly replenish, 250 gallons of DEF. In addition, room would have to be found for the dosing cabinet, which houses the pumps and controls and is where DEF is added to engine exhaust, and the "Clean Emissions Module," which contains the catalytic converter.

Above, brand new Progress Rail EMD F125 passenger locomotive, Muncie, IN, July 2016. Image Credit: Progress Rail, A Caterpillar Company.

On the right, Los Angeles-based Metrolink factsheet, explaining the benefits of its new Tier 4 F125 locomotives. Image Credit: Metrolink - Copyright © 2016 Southern California Regional Rail Authority.

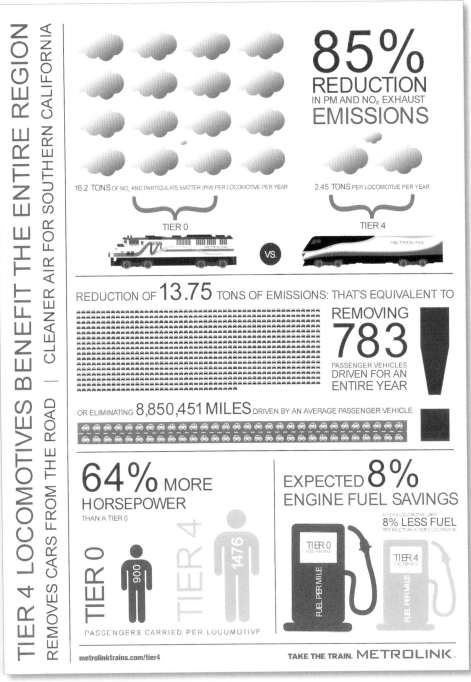

5.4 Impact of Tier 4 on Fuel Economy

Locomotive diesel engine experts will tell you, "when you improve NOx, everything else gets worse," and "each improvement in emissions has made locomotive engines less efficient."[195] This dilemma has been true from the beginning of EPA locomotive emissions control, when NOx reduction was accomplished by retarding engine timing, to current Tier 4 strategies that rely on EGR and SCR. The *Railroad and Locomotive Technology Roadmap* described it this way: "Unfortunately, most techniques for reducing NOx also decrease the fuel efficiency of the engine and raise PM emissions."[196] This authoritative report specifically stated that both basic approaches for further reducing NOx—EGR and SCR aftertreatment—would have negative impacts on fuel economy. As indicated in the sidebar, these concerns were also echoed in the manufacturers' arguments to the EPA during the Tier 4 rulemaking process.

GE and EMD have declined to publish data on how much of an efficiency loss occurred with Tier 4 or whether compensating efficiency gains in other areas were sufficient to maintain or improve overall efficiency when compared to their Tier 3

GE, EMD, and EPA Tier 4 Negotiating History

In 2008, during the Tier 4 rulemaking process, both GE and EMD brought to EPA's attention that further reductions of NOx from locomotives would have a substantial energy penalty. However, after agreeing on this point, the comments of these two manufacturers diverged.[195]

GE referenced the 2007 *Massachusetts v. EPA* Supreme Court decision (which compelled the EPA to regulate greenhouse gas emissions) and argued that the Tier 4 NOx reduction requirement should not be so stringent that it increased fuel consumption, and therefore greenhouse gas emissions—since the latter would be counter to the Supreme Court decision.

EMD, on the other hand, requested the opposite, i.e., that the EPA ignore GHG emissions in the Tier 4 rulemaking, because requiring reductions in GHG emissions would further complicate the already difficult task of achieving proposed Tier 4 NOx emissions reductions. EMD also pointed out that urea aftertreatment produces carbon dioxide emissions, which, if counted, would further complicate the task of simultaneously reducing both NOx and GHG emissions.

EPA rejected parts of both of these arguments. In response to GE, it stuck with a stringent NOx reduction standard. And in response to EMD, the EPA concluded that the NOx standard would not increase fuel consumption (and therefore GHG emissions) by very much. In fact, EPA optimistically said it would be by less than 1%. EPA further stated that this small increase in fuel consumption could be more than offset by a variety of already-identified diesel engine energy conservation strategies.

Thus, while demonstrating a concern for energy efficiency and GHG emissions, the EPA did not directly address climate change impact in Tier 4 rulemaking. This is likely to change at some point in the future when federal government policy on climate change returns to being rational and science-based.

It should be noted that acid rain-producing sulfur dioxide emissions (SOx) were also not addressed in this rulemaking. SOx emissions had already been addressed by another EPA program that mandated the use of low sulfur diesel fuel (500 ppm) by 2007, and ultra-low sulfur diesel fuel (15 ppm) by 2012.

locomotives. The goal of exceeding Tier 3 locomotive efficiency was laudable but achieving that goal would be very difficult.

Not only does EGR reduce combustion efficiency by mixing exhaust gases with fuel and intake air (diluting fuel and oxygen concentrations with combustion products), but it also reduces efficiency by lowering combustion temperature. Additionally, EGR can create back pressure on the exhaust manifold, which makes engine pistons work harder to eject exhaust gases. Finally, EGR requires additional cooling, which adds cooling fan horsepower and auxiliary load.

The efficiency impact of urea aftertreatment is less clear. While there are small energy penalties associated with pumping DEF and the back pressure that aftertreatment imposes on engine exhaust, the aftertreatment approach could have an energy efficiency benefit if the use of aftertreatment allows engine timing to be advanced. In that way a 3% to 5% increase in the use of DEF, could produce a 2% improvement in fuel economy.[198] However, a comprehensive analysis of the urea aftertreatment option should treat DEF as a fuel since its use in the SCR process produces carbon dioxide emissions (in nearly the same volume as NOx reductions) and urea is made with ammonia, which is manufactured with natural gas. Including these additional "fuel" inputs in the equation would mitigate the efficiency/GHG advantage sometimes ascribed to aftertreatment over EGR.

Note also that new Tier 4 passenger locomotives, which use urea aftertreatment, also use lightweight high-speed diesel engines. These tend to be less efficient than the heavier weight medium-speed diesels used by freight locomotives, but they produce less NOx because operating at higher revolutions per minute reduces the time available for combustion.

So, what's the answer? Has Tier 4 reduced locomotive efficiency? While in the past GE and EMD have been able to overcome energy penalties associated with stricter emissions standards and produce new locomotives that were both cleaner and more efficient, they do not appear to have been successful with Tier 4. An overall decrease in efficiency seems likely, but it's hard to document in the absence of publicly verifiable independent test results or transparency by the manufacturers. The author has been told by experts in the field that EGR has reduced Tier 4 diesel engine BSFC significantly, and that overall locomotive efficiency has declined compared

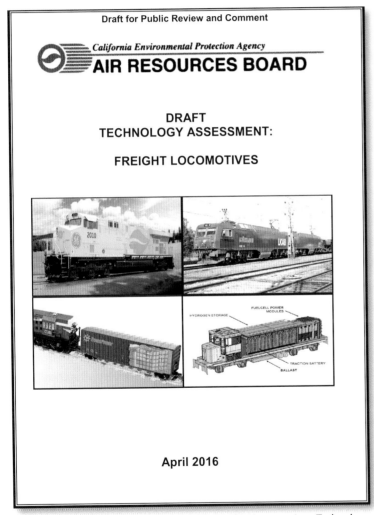

California Air Resources Board Draft Freight Locomotive Technology Assessment Report, April 2016. Image Credit: California Air Resources Board.

to Tier 3. A "3% efficiency loss" is most often given, though one builder's sales representative quote was "less than 1%."

This lost ground is significant, but it may have a smaller fuel consumption impact than imagined given the current trend to defer new locomotive purchases in favor of rebuilding older locomotives. The latter option incorporates new electronics and other components and uses less-encumbered engines of earlier design to meet lower emissions standards than Tier 4. These locomotives are at least more energy efficient than the older locomotives from which they were made.

5.5 Tier 5 Emissions Standards?

As of this writing, the EPA has not announced plans to develop Tier 5 locomotive emissions standards that would require even deeper reductions in NOx and other "criteria" pollutants, and probable first-time cuts in locomotive greenhouse gas emissions. Consideration of a national EPA Tier 5 seems very unlikely during the Trump Administration. Meanwhile, California has indicated a strong interest in Tier 5.

On July 17, 2015, California Governor Edmund G. "Jerry" Brown, Jr., issued an executive order on "Sustainable Freight Transport."[199] This order, among other things, called for improved freight railroad energy efficiency and a transition to zero-emissions locomotive technologies while increasing railroad competitiveness. The executive order's action plan[200] is being developed by the California Air Resources Board in conjunction with other state agencies. It reflects California's continuing concerns about air quality and supports its commitment to reduce greenhouse gas emissions 40% below 1990 levels by 2030. One portion of that plan, entitled "CARB Draft Technology Assessment: Freight Locomotives" (CARB Report), specifically examines freight locomotives.

The CARB Report calls for a national Tier 5 diesel-electric locomotive emissions standard that mandates an additional 85% reduction of NOx and 75% reduction of particulate matter compared to Tier 4.[201] Plus, for the first time, locomotive GHG emissions would be regulated with a mandate that they be reduced by as much as 25%.

Proposed California Tier 5 Locomotive Emissions Standard

Emission Levels	NOx (g/bhp-hr)	PM (g/bhp-hr)	GHG Reductions (relative to Tier 4)	Cost per Locomotive (Million $)	Potential Control Technology
a. Tier 4 Standard	**1.3**	**0.03**	0%	$3.0	EGR, turbos, cooling
—In-Use	1.0	0.015	N/A	N/A	
b. Tier 4 with Aftertreatment	**0.3**	**0.01**	0%	$3.25	Compact SCR and DOC
—In-Use	0.2	<0.0075	N/A	N/A	
c. Tier 4 with Aftertreatment and On-Board Batteries	**0.2**	**0.0075**	10-25%	$4.0	Compact SCR and DOC + On-Board Batteries
—In-Use	0.15	0.006	10-25%	N/A	

Table ES-5. CARB Freight Locomotive Technology Report. The "In-Use" row at the bottom further reduces emissions to levels CARB determined would be needed to ensure compliance at the end of the design life of the locomotive. "DOC" refers to diesel oxidation catalyst technology. Image Credit: California Air Resources Board

According to CARB staff, locomotive manufacturers could achieve these combined emissions reductions by:

1. Installing advanced aftertreatment on locomotives that have achieved Tier 4 emissions levels without aftertreatment (this option is "b" on the preceding page)

2. Adding hybrid/battery technology to option "b" (this is option "c" on the preceeding page)

How technically feasible is CARB's Tier 5 locomotive emissions standard? One way to answer that question is to observe that since 2010 heavy-duty trucks have had to meet emissions standards that require NOx and particulate emissions to be 85% and 66% below current Tier 4 locomotive standards, respectively.

CARB acknowledged that finding available space within a Tier 4 freight locomotive for aftertreatment and a hybrid battery system would be challenging, but it assumed technological advances will solve this problem by making both systems smaller and more capable.[202] It specifically assumed that the hybrid battery would be similar to that used in the GE prototype hybrid locomotive, though 50% more effective.[203] CARB estimated conservatively that these improvements would increase the cost of a $3 million conventional Tier 4 locomotive by 33%, to $4 million.[204] (Note that other industry sources give $3.5 million as the cost of a Tier 4 freight locomotive.)

However, in comments submitted to CARB on July 13, 2016, GE questioned CARB's Tier 5 by pointing out difficulties and drawbacks of applying aftertreatment to their existing locomotives. Additionally, GE noted that the development of hybrid diesel-electric locomotives was in its infancy and should not be considered a near-market technology.[205] (See discussion of hybrid locomotive technology in the next chapter.)

GE's concerns notwithstanding, locomotive builders may eventually meet Tier 5 emissions standards with advanced SCR aftertreatment systems, abandoning EGR entirely. The abandonment of EGR would allow efficiency gains and follow the pattern of the trucking industry (which also abandoned EGR in favor of SCR to meet higher emissions standards). As such, Tier 5 would impose on the freight railroads the cost of DEF and its supply infrastructure. Depending on a number of variables, those costs could be offset by fuel savings. However, irrespective of technological approach, if the cost premium of Tier 5 emissions reduction was excessive, it could reduce the sale of new locomotives and further incentivize railroads to rebuild their existing stock. Thus, NOx and particulates emissions gains might be less than anticipated.

On April 13, 2017, California formally requested that the EPA begin a rulemaking process for Tier 5 locomotive emissions standards.[206] As of May 2018, there has been no response from the EPA.

Somewhere on the Amtrak Northeast Corridor after 2021. © Alstom.

CHAPTER SIX

Future Directions for Diesel-Electric Locomotives

This chapter more clearly represents the author's environmental perspective – specifically, his conviction that we must take quick, decisive, and effective steps to address climate change before it spirals further out of control. This view, while personally held, is supported overwhelmingly by the scientific community. The vast majority of climatologists affirm that climate change is real, caused by human activity, and is accelerating with potentially devastating consequences. Readers who want more information about climate change may obtain it from the excellent books, reports, and resources listed in the "climate change" section of this book's reading list. A list of consequences of climate change is available in the preface.

6.1 The Challenge of Climate Change

The analysis presented here is a consensus view[207] shared by virtually every credible scientific organization in the world, led by the United Nations' Inter-governmental Panel on Climate Change (IPCC), whose scientific work involves thousands of scientists and over 120 governments.[208] The IPCC determined over 10 years ago that industrialized countries must reduce their greenhouse gas emissions by 80% by 2050 while the world as a whole reduces its GHG emissions by 50% in order to prevent runaway catastrophic climate change. As evidence of accelerating anthropogenic (human-caused) climate change has increased since then, many climatologists and advocates for vulnerable nations have begun to call for even deeper cuts in emissions. Meanwhile, 2016 was the hottest year on the historical record (since 1880), and 16 of the 17 hottest years on this record have all taken place since 2001.

Climate change is occurring because the concentrations of GHGs in the atmosphere are increasing due to human activity. These gases—carbon dioxide, methane, nitrous oxide, ozone, and chlorofluorocarbons—act like a transparent thermal blanket for the earth. They allow light from the sun to strike the earth and warm it but then block and absorb some of that energy as it radiates back into space. Some of that absorbed heat is then re-radiated back toward the earth, warming it again, and in the process slightly increasing the temperature of the earth's atmosphere. This "Greenhouse Effect" is good because it has warmed the earth's surface to a life-supporting average of 59 degrees. But, as concentrations of GHGs have increased because of human activities (principally burning fossil fuels and destroying forests), the atmosphere has continued to warm up—now at an increasing rate. The pre-industrial atmospheric concentration of carbon dioxide was 280 ppm. In the beginning of 2018, the level was 410 ppm and climbing.

Scientists have known about the danger of climate change for many decades but their warnings have gone unheeded. The problem has been called an *inconvenient truth* because the global economy is so heavily dependent on fossil fuels and there is no

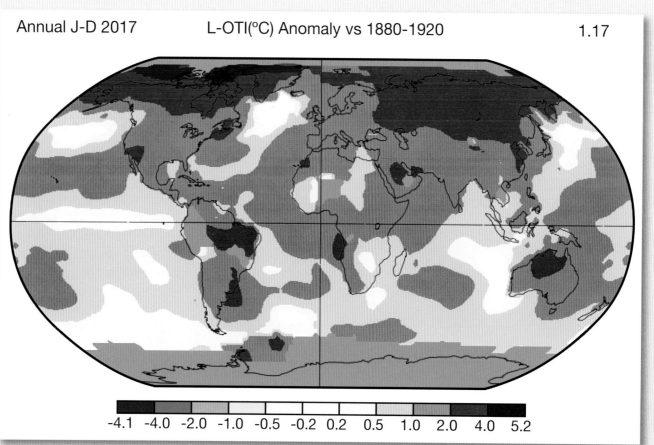

-4.1 -4.0 -2.0 -1.0 -0.5 -0.2 0.2   0.5   1.0   2.0   4.0   5.2

Map showing 2017 global average temperature rise in degrees Centigrade compared to 1880-1920 average. Note severe warming already occurring in the northern latitudes especially in the Arctic region. The gray area indicates insufficient data. The IPCC has warned that as much as 8°F average warming is possible by the end of the century if we follow a "business as usual" path and fail to sufficiently reduce GHG emissions. The latest IPCC report calls for zero GHG emissions by 2050. Image Credit: NASA.

quick fix. The only way to reset the global thermostat is to transition away from coal, oil and natural gas—useful fuels that have improved our lives in many ways—and replace them with non-GHG-emitting renewable energy sources such as solar, wind, biomass, geothermal, and hydroelectricity. Nuclear power could also be part of a clean energy future if nuclear plants of advanced design and a reformed nuclear fuel cycle were able to solve problems associated with safety, waste, diversion, and proliferation.

A transition to low carbon and no carbon emissions energy sources is now underway but needs to be greatly accelerated. Energy conservation is critically important to this transition. Without a much more serious commitment to conservation, it's doubtful that low- and no-carbon energy resources could meet our energy demand.

(*Note:* In this discussion, the author uses carbon, carbon dioxide, and GHG emissions more or less interchangeably.)

While the threat of climate change requires an international response, all sectors of the U.S. economy must be part of a national effort to reduce GHG emissions by at least 80% by 2050. That includes the railroad industry whose diesel-electric locomotives must become much more energy efficient and operate with substantially reduced greenhouse gas emissions.

The chart below outlines the energy efficiency and energy source options discussed in this chapter.

An examination of future directions for diesel-electric locomotives is an exercise in pro and con, plus and minus. There is no free lunch. There are no easy answers. There are challenges and costs associated with all options. In fact, one recent study sponsored by the Federal Railroad Administration concluded that the costs of all alternatives to conventional diesel fuel-powered diesel-electric locomotives were too high to be cost-justified.[209] Our emphasis here, however, is to prioritize reducing GHG emissions while still looking for affordable solutions.

Low- and No-Carbon Emissions Options for Diesel-Electric Locomotives

	Technological Readiness	GHG Reduction Potential	Cost	Comments
Efficiency				
Waste Heat Recovery	Close	Significant	Medium	Multiple technologies
Hybrid Locomotive	Not Yet	Significant	High	GE project incomplete
Energy Storage Tenders/ZEBLs	Not Yet	Significant	High	Battery issues
Alternate Fuels				
Natural Gas	Ready	None/Minus	High	Methane slip/Infrastructure $
Fischer-Tropsch Process	Ready	Modest	High	Feedstock choice/efficiency
Biofuels	Close	Significant	High	NOx/infrastructure $
Hydrogen Fuel Cell	Close	Significant	High	Carbon neutral fuel source issue Primarily passenger appl.
Electrification				
Full – Catenary	Ready	Significant	Very High	New infrastructure/motive power
Partial	Ready	Significant	High	Dual source locos, tenders, ZEBLs
Buy/Build Green Power	Ready	Depends	Low/Med	Additionality & market problems
Carbon Credits				
	Ready	Depends	Medium	Additionality & market problems

6.2 Waste Heat Energy Recovery

If a diesel-electric locomotive's diesel engine is operating at 40% efficiency, then 60% of the energy in its diesel fuel is not converted into useful work and is wasted. This energy exits the engine as waste heat from its exhaust stack and radiators—where the latter reject waste heat from (a) the engine block, (b) the oil cooler, (c) combustion air intercoolers/aftercoolers, and, in the Tier 4 locomotive, (d) the EGR cooler. Additionally, there are radiant heat losses from the engine block and other hot surfaces, although such losses represent a very small percentage of engine heat loss.

In pre-Tier 4 locomotives, the waste heat rejected by the exhaust stack and radiators was roughly equal. But in Tier 4 locomotives, the balance shifted toward radiator-rejected heat because the EGR system removes exhaust gases from the locomotive stack. The cooling of these gases is then accomplished by the locomotive's engine cooling system.

There are a variety of ways to recover and reuse locomotive prime mover waste heat. Foremost among them is the turbocharger, which was discussed earlier. Using multiple turbochargers arranged in series creates a compounding effect, which has allowed locomotive builders to recover

and use additional amounts of waste exhaust heat in their Tier 4 locomotives.

Eventually, it may be possible to convert heat energy in the locomotive exhaust directly to electricity using semiconductor *thermoelectric generators* (TEGs). But this technology is not yet well developed. More promising at this time is the organic Rankine cycle.

An organic Rankine cycle Waste Heat Recovery System (WHRS) uses a working fluid such as water or a refrigerant to recover energy typically from low-temperature heat. The working fluid boils when it is heated in a heat exchanger located in the engine exhaust stack or cooling system. The working fluid, now a hot gas, then turns a turbine connected to a generator that produces electricity. This electricity can then be used to power locomotive auxiliaries or provide additional traction horsepower. An air-cooled condenser returns the working fluid to liquid form so that it can be continuously reused in this closed loop cycle, capturing waste heat and generating electricity with it.

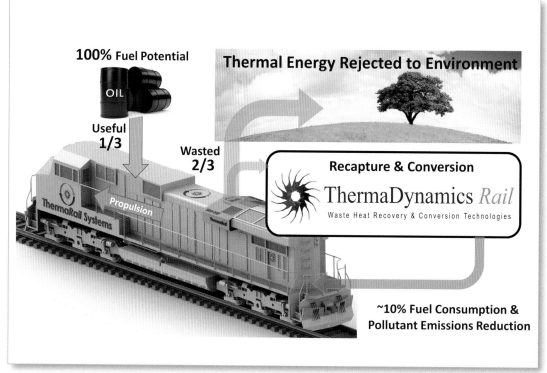

Energy recovery from locomotive exhaust stream. The diagram assumes a locomotive operating at 33% efficiency prior to recovering 10% of its waste heat. Image Credit: ThermaDynamics Rail.

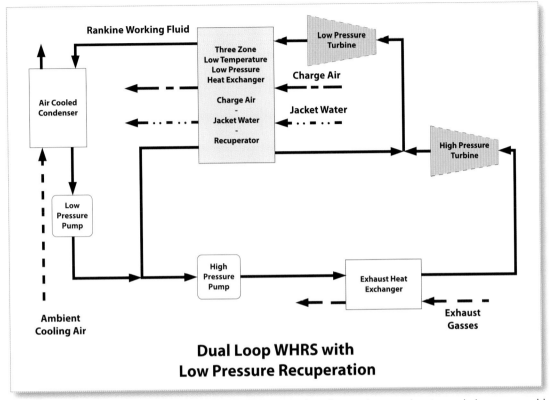

Dual Loop WHRS with Low Pressure Recuperation

Schematic diagram depicting how locomotive engine waste heat from exhaust and water-cooled sources could be recovered. The high- and low-pressure turbines would power generators to produce additional electricity for auxiliary and traction purposes. "Low pressure recuperation" refers to additional heat recovery from the exhaust of the low-pressure turbine—instead of having this heat rejected into the atmosphere by the condenser. Various other designs are possible to optimize energy recovery. Image Credit: Dave Cook at Rail Propulsion Systems.

Theoretically, Rankine cycle processes can produce a meaningful increase in locomotive energy efficiency. Consider this simple "back of the envelope" calculation:

Assumptions:

- Diesel fuel input is 100 energy units.
- Diesel engine efficiency is 40% (40 energy units) and energy waste is 60% (60 energy units).
- The exhaust system and coolant system are equally productive of waste heat (30 energy units each).
- Radiant heat loss from the engine block and other hot surfaces is a very small percentage and is ignored.

- 40% of exhaust system waste heat is recovered by heat exchanger.
- 30% of the coolant system waste heat is recovered by heat exchanger.
- 25% of recovered heat is converted to electricity by Rankine cycle(s).

30 units of energy waste from the exhaust system x 0.40 eff. x 0.25 eff. = 3.00 units of energy put to work

30 units of energy waste from the coolant system x 0.30 eff. x 0.25 eff. = 2.25 units of energy put to work

Total waste heat recovery = 5.25 energy units

In this hypothetical scenario, the recovery of diesel engine exhaust and coolant waste heat leads to a 5.25 energy unit improvement—minus, of course, the parasitic loads associated with pumps, fans, and compressors where required. Subtracting one energy unit for these parasitic losses still leaves an efficiency gain of 4.25 energy units, bringing our hypothetical 40% efficient prime mover to 44.25% efficiency. If this kind of efficiency improvement was achieved and applied to operating locomotives (including older retrofit locomotives), it could be game-changing for the railroads. It would, after all, represent an efficiency improvement close to that claimed for energy management systems like Trip Optimizer and LEADER.

If locomotive Waste Heat Recovery Systems have so much potential, why haven't they been implemented? Here are some barriers:

Heat exchanger design—The exhaust stack heat exchanger must be able to reliably handle thermal cycling with maximum 700°F -1,000°F temperatures, and effectively recover heat without imposing back pressure on the engine.

Working fluid limitations—More efficient working fluids are needed to adequately capture stack energy waste without temperature-related decomposition.

Space constraints—There is not much available space inside most locomotives for additional heat exchangers and other equipment.

Complexity and maintenance—Heat recovery would add an additional layer of complexity that could impact maintenance, and, potentially, reliability.

Warranty issues (if retrofit)—Locomotive manufacturers may void warranties—a problem only for newer locomotives.

Cost and payback—Because railroads don't want to add cost to an already expensive asset, the payback of a WHRS must be quick.

These barriers have impeded development and implementation of WHRS. However, while GE and EMD have shown little interest in waste heat recovery, others are exploring this energy saving opportunity.

ThermaDynamics Rail of Manassas Park, Virginia, is developing waste heat recovery retrofit systems designed to recover 10% to 17% of locomotive waste heat and produce in excess of $110,000 in annual fuel savings – based on 2016 diesel fuel prices and the EPA locomotive duty cycle.[210]

The company's exhaust waste energy heat recovery system relies on a high-pressure heat exchanger (HiPHEX) to recover heat from the locomotive's exhaust. The heat exchanger is uniquely designed to create minimal exhaust back pressure and withstand the expansion and contraction cycles caused by changes in exhaust temperatures. These cycles occur during "transients" when load and throttle settings change.

According to company lead scientist and manager Claudio Filippone, ThermaDynamics Rail's goal is to create a locomotive WHRS that fits inside existing locomotive "real estate" (available space) and can be installed "non-invasively" overnight during regular locomotive maintenance operations with minimal inconvenience. Filippone projects that his heat recovery system could be ready for commercial sale as early as 2018 and have an attractive 3-year payback. ThermaDynamics is also developing an organic Rankine cycle-based heat recovery system to recover cooling water system waste heat. This additional heat recovery would increase annual fuel savings and potentially shorten the payback period.

ThermaDynamics Rail's high-pressure heat exchanger designed to be installed on the turbocharger outlet of a 16-cylinder locomotive engine. It captures waste heat from exhaust gases, transferring this energy to a working fluid (in this case, water) circulating in a closed-loop of a Rankine power generation cycle. Image Credit: ThermaDynamics Rail.

Waste heat recovery power generation cycles using other working fluids, such as supercritical carbon dioxide (sCO_2), may also show promise for locomotives in the future. Supercritical fluids are substances that do not have a defined phase change between liquid and vapor and can exhibit characteristics of both. This can be thermodynamically advantageous, depending on the application.

A heat recovery cycle using sCO$_2$ would look much the same as a conventional Rankine cycle. A heat exchanger inserted into the locomotive exhaust stack would capture and transfer exhaust heat to the sCO$_2$. The heated supercritical fluid would then turn a turbine connected to a generator that produces electricity for locomotive functions. The cycle would also include a heat rejection heat exchanger and a low temperature compressor. A recuperative heat exchanger is used to re-cover heat from the turbine exhaust in order to preheat the working fluid before it entered the heat exchanger in the stack. This heat recovery is important to the sCO$_2$ power cycle to keep thermodynamic efficiencies reasonable.

To be successful, sCO$_2$ waste heat recovery systems would

Metra trains in Chicago, at Roosevelt Bridge, and Evanston, IL, station. Like many commuter railroads, the Metra is push-pull operation. While outgoing trains are pulled, incoming ones are pushed. If the hybrid locomotive technology discussed in the following pages were applied to the Metra, the energy recovered during frequent station stops could be put to work propelling the train between stations. Photo Credits: Judie Simpson.

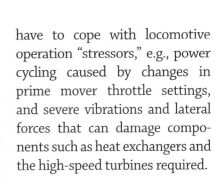

have to cope with locomotive operation "stressors," e.g., power cycling caused by changes in prime mover throttle settings, and severe vibrations and lateral forces that can damage components such as heat exchangers and the high-speed turbines required.

6.3 Hybrid Locomotives, Battery Tenders, ZEBLs

A hybrid vehicle is powered by both an internal combustion engine and one or more electric motors, and it uses regenerative braking and a storage battery to maximize energy efficiency. Regenerative braking is similar to dynamic braking (with electric motors functioning as generators), but the electricity that's generated is recovered and used for motive power instead of being wasted to the atmosphere as heat released in dynamic braking grids. While electric locomotives can discharge recovered braking energy into catenary lines (for use by other electric locomotives), diesel-electric locomotives need batteries or other portable storage media to capture this braking energy.[211]

Some automobiles with hybrid drive systems have average fuel economies greater than 50 miles per gallon. Although hybrid cars may seem to be a relatively new idea, they appeared as early as the 1910s (see Section 1.3). These "dual power" cars may look primitive now and surely were technologically deficient by present standards, but they had functional hybrid drives that included regenerative braking and battery storage. Over 100 years later, it should be possible to incorporate this technology into diesel-electric locomotives.

The *Railroad and Locomotive Technology Roadmap* noted that a slow freight train operating through the mountains would dissipate as much as 64% of its kinetic energy as heat in its dynamic braking grids.[212] That "braking energy ratio" (given here as a percentage) represents the upper end of what could be recovered by a locomotive regenerative braking system. After factoring in battery and electric traction motor efficiencies, actual maximum recoverable kinetic energy by a hybrid diesel-electric locomotive might be more like 55%—though, of course, recovery goes to zero once the battery is full. Thus, battery capacity matters, though it must be balanced against other factors, e.g., size, weight, and cost.

Diesel-electric locomotives pulling long-distance freight or intercity passenger trains would have a much lower braking energy ratio, and therefore would not be prime candidates for hybrid diesel-electric locomotive systems. In contrast, quick commuter rail trains with frequent stops could be good candidates. Examples abound and include Chicago's Metra, Southern California's Metrolink, Philadelphia's SEPTA, and New York City's Long Island and Metro North Railroads.

Green Goat Hybrid Locomotive
Approximately 55 Green Goat hybrid switcher locomotives were manufactured by Railpower Technologies after they were introduced in 2004. Many were purchased and used by Union Pacific.[213] These locomotives are no longer for sale[214] and all were reported retired by Union Pacific as of the end of 2014.

Green Goat locomotives were typically built on the frames and trucks of recycled EMD GP9 locomotives.[215] They used a 300 hp diesel engine to charge a bank of batteries that were capable of delivering 2,000 hp and a respectable 80,000 pounds of tractive effort. Since the diesel engine could operate at a constant speed, it operated efficiently and relatively cleanly. Railpower claimed Green Goats achieved up to 50% energy savings over conventional diesel-electric switchers.

This "battery dominant" hybrid, however, was essentially an electric locomotive with a small engine to recharge batteries. While this concept might work well for a switcher locomotive, it could be argued that Green Goats were not true hybrids because they did not use regenerative braking to recover energy for battery recharging.[216] Sales were hampered when one early-generation Green Goat caught fire. While yard switchers are constantly starting and stopping, they may not be good hybrid candidates because they operate at low speeds and therefore have low amounts of kinetic energy to recover.

GE Hybrid Locomotive

With much fanfare, GE unveiled a prototype hybrid road locomotive in front of Los Angeles' Union Station on May 24, 2007.[217] Accompanying the rollout was GE's announcement that it had spent 5 years and $250 million developing the locomotive and its batteries.

GE's hybrid locomotive project received federal funding from the previously mentioned U.S. Department of Energy's 21st Century Locomotive Technology program (2003-2010). DOE documents indicate these R&D tasks specific to the development of hybrid locomotive:

- Develop advanced batteries and a battery energy management system to be used in a hybrid diesel-electric locomotive
- Develop an advanced locomotive operating system for a hybrid locomotive, i.e., a Hybrid Trip Optimizer
- Develop, build, and test a prototype hybrid locomotive

BNSF and Union Pacific were said to be on the project's advisory board.

GE used a modified 4,400 THP ES44AC as a test bed for the project. Company reports characterized the prototype as though it had achieved these energy and environmental milestones,[218] presumably in comparison to GE Tier 2 locomotives:

- 10% to 15% energy savings (32,000+ gallons of diesel fuel/year)
- 10% emissions reduction
- Availability of an additional 1,750 horsepower for short durations

Significantly, the beautifully painted hybrid locomotive displayed in various locations around the country was numbered 2010—the date when GE apparently hoped to start selling production models. But nearly a decade has passed since then, and there have been no completed production models. Neither GE nor the DOE has published test results on the performance of the hybrid prototype or explained why the project was discontinued. Mention of the hybrid locomotive disappeared from the GE website and public relations materials.

The GE hybrid diesel-electric locomotive prototype at Belt Railway Company of Chicago Clearing Yard, Bedford Park, IL. September 16, 2008. Photo Credit: Sean Graham-White.

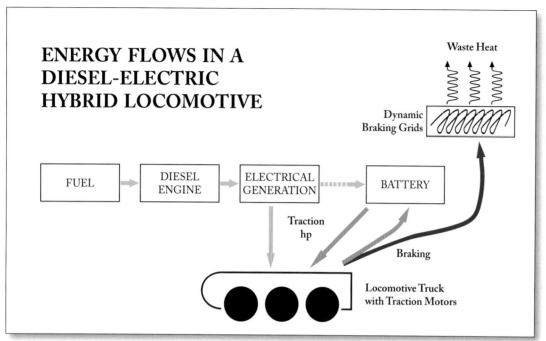

ENERGY FLOWS IN A DIESEL-ELECTRIC HYBRID LOCOMOTIVE

Image Credit: Robert Hochberg.

Market Readiness, Cost, and Payback Issues. So, what happened? In 2010, a GE spokesman told a reporter at the *Erie Times-News* newspaper (Erie, PA) that the hybrid locomotive would be ready when buyers were ready, and that GE could have one available for sale by the end of the year if a railroad ordered one.[219] The GE spokesman further noted that the company's hybrid locomotive would come at a premium price, but he did not indicate what that price would be or how quickly it could be paid back by fuel savings. He added, "Everybody wants to increase fuel efficiency and

lower emissions, and hybrid technology will help meet these requirements. But with oil prices as they are today, there is no compelling reason for a railroad customer to switch."[220]

Diesel fuel prices were $2.26/gallon in 2010, when GE's hybrid locomotive was supposed to be in production. Prices rose to $3.17/gallon in 2012 and dropped to $1.79/gallon in 2015. These are all relatively low prices, much too low to make hybrid locomotives economically attractive and marketable. While railroads, like other consumers, want low cost energy,

higher energy prices are clearly needed to stimulate genuine interest in the kind of energy conservation and efficiency improvement required to address climate change. A carbon tax or fee[221] could raise energy prices sufficiently to prompt new energy saving technologies like hybrid locomotives. But this kind of financial incentive did not exist in 2010, and still does not exist as this book is being completed in 2018. (Note: The financial burden of carbon fees can be minimized through a "carbon fee and dividend" program which strives to be revenue neutral by returning average fees to energy users.[222])

The CARB Report stated that, as of April 2016, the GE hybrid locomotive needed both mid- and full-scale demonstrations in locomotive field service (which would take at least 2 years) before it could be ready for commercial production. CARB further stated that "GE does not currently have plans to put this technology into production."[223] Concerning hybrid locomotive payback, CARB estimated that the payback for a hybrid diesel-electric locomotive would be 11 years, assuming a $1 million premium price, $90,000 in annual fuel cost savings, and constant 2016 diesel fuel prices.[224] A life cycle analysis would factor in maintenance and battery replacement costs, as well as other factors, lengthening the payback period.

In response, on June 15, 2016, GE commented on the CARB Report, explaining that the battery system in its prototype hybrid locomotive was only one quarter the size needed to produce the 10% fuel savings promised.[225] Thus, while demonstrating the feasibility of a hybrid diesel-electric locomotive, GE had not actually achieved the 10% fuel savings. GE noted that additional batteries would have added significant weight to the locomotive, and their test bed locomotive was already at maximum allowable weight of 432,000 pounds. GE also stated that at current battery prices the cost of a hybrid locomotive with claimed performance abilities would exceed CARB's estimates, further extending the payback period of the locomotive—making it even more uneconomical.

The TRL/Ricardo UK study, previously mentioned in Section 2.5, examined a "mild hybrid" option for UK passenger locomotives bundled with other efficiency items, a urea SCR aftertreatment system, and new diesel engines.[226] The authors of the study concluded that while this larger package could achieve a 34% to 42% fuel savings, it would have a 20-year payback. Mild hybrid capability (defined as not having significant energy storage) was said to be able to achieve a 10% to 15% energy saving when bundled with Automatic Engine Start Stop anti-idle technology.[227] The study, incidentally, includes excellent assessments of many diesel-electric locomotive energy saving technologies.

Battery Development Issues. GE's hybrid locomotive program appears to have stalled because of the difficulty GE had developing suitable batteries. A 1.5-megawatt battery pack was planned. When charged with electricity from regenerative braking, this battery pack should have been able to deliver an additional 1,750 to 2,000 hp to the locomotive's electric traction motors. But concept is one thing; a finished product is another.

The wide-ranging *Railroad and Locomotive Technology Roadmap*[228] identified essential hybrid locomotive battery attributes. These batteries would need to have:

- Very large energy storage capacity (kilowatt-hours)
- Very high energy density (kilowatt-hours per unit of weight and volume)
- Very high power densities (kilowatts per unit of weight and volume)
- Very rapid charge and discharge times
- Deep discharge capacity
- Long lifespan
- High energy efficiency

The battery in a (non-plug-in) hybrid automobile is designed to store a modest amount of electricity produced by regenerative braking, and to provide short bursts of power before needing recharging. In contrast, batteries needed for a successful hybrid diesel-electric locomotive would have to be able to store huge amounts

GE's hybrid prototype apparently still undergoing testing on the GE test track, Erie, PA. October 25, 2013. Photo Credit: Stephen Grekulak.

of energy and receive and discharge that energy quickly—without degradation. These batteries would also have to fit within the confines of an already packed locomotive and not be so heavy that other locomotive energy systems had to be downsized. Quite a tall order.

While General Electric selected sodium nickel chloride batteries for its hybrid prototype locomotive tests,[229] its battery of choice for a hybrid locomotive was a "high energy density, sodium-based battery."[230] GE planned to make its hybrid locomotive batteries at a new $105 million battery plant in Schenectady, New York, beginning in 2011, but by 2016 the plant was closed.[231] GE cited battery costs and market pressures as its motivation for closing the plant.

GE's decision not to complete commercialization of its hybrid diesel-electric locomotive probably was also influenced by the company's resource-intensive effort to design and build a Tier 4 locomotive by the EPA's January 1, 2015, deadline.[232] (The hybrid was not designed to be Tier 4 compliant.)

Hybrid Barriers. In sum, while the introduction of a much more energy efficient diesel-electric locomotive like the hybrid is critically important, its arrival has been delayed by numerous factors, including:

- Low oil prices that decrease the dollar value of energy saving
- Insufficient battery development
- Projected high battery cost
- Projected high purchase price of a hybrid locomotive
- Projected long payback for prospective buyers
- GE's commitment of resources to other projects

To this list, we can add:

- Absence of government incentives, subsidies, mandates, or carbon tax/fee
- Soft and unpredictable locomotive sales market

These are not insurmountable odds but overcoming them will be challenging. The hybrid freight locomotive will be revisited, and its eventual production and use is a given.

Already hybrid locomotive technology is being applied to passenger trains, e.g., Bombardier's Mitrac Hybrid dual-power locomotive with its optional energy saving system capable of storing regenerative braking energy when the locomotive is not operating under catenary.[233] The hybrid concept applied to freight hauling might actively progress with battery tenders after further battery development.

Energy Storage Tenders for Hybrid Locomotives

Diesel-electric locomotives are ready-made for hybrid operation because they have internal combustion engines, electric traction motors, and the potential for regenerative braking. All that is needed is energy storage. If finding room for enough energy storage on-board the locomotive is a road block, space could be found in a separate car or tender that is attached to the locomotive.

Energy storage tenders can carry large banks of electrochemical batteries (e.g., lead-acid, nickel-metal hydride, sodium nickel chloride, lithium ion, etc.). Alternately, they can house ultracapacitors, electric flywheels, compressed air storage, fuel cell hydrogen storage, or superconducting magnetic energy storage.

Each of these storage media has its own technical challenges.[234] Moreover, converting from one type of energy to another involves inevitable energy losses. In batteries, energy is lost when electrical energy is converted to chemical energy (which occurs during battery charging), and it's lost again going in the reverse direction (when extracting electricity from the battery). That said, modern lithium ion battery efficiency is very high; these modern power cells have low DC resistance and can have efficiencies as high as 97% even at relatively high currents.

Norfolk Southern's 1,500 hp battery-powered switcher crossing the Roanoke River on its way to Roadway Material Yard, Roanoke, VA. Photo Credit: © Norfolk Southern Corp.

Battery tenders present opportunities beyond simply holding stored regenerative braking energy. For example, these tenders could be charged from an on-board diesel engine (like the Green Goat switcher) or hydrogen fuel cell. Or they could be charged from external electrical power sources via a third rail, overhead wires, or wireless power transfer systems. At stations and terminals, idle battery tenders could be plugged in for recharge. However, none of these charging options would be zero carbon emissions unless the charging power came from a carbon-free emissions energy source.

Battery tenders that have electric motors can be called *hybrid booster locomotives*. These can both store regenerative braking energy and add tractive effort as needed.[235] For passenger trains, which travel at higher speeds than freights, tenders can also provide a useful boost in horsepower to accelerate and achieve high speeds more quickly.

Since 2007, Norfolk Southern has been developing an all-electric battery-powered locomotive. The second generation uses lead-carbon hybrid batteries (in-

stead of lead acid batteries) and a more advanced battery management system.[236] It is easy to imagine this unique all-electric switcher evolving into a motorized battery tender positioned between two locomotives, permitting regenerative braking energy recovery. Norfolk Southern holds a number of patents pertaining to battery locomotives, including one for a hybrid electric line-haul locomotive.[237]

In 2012, Transpower,[238] a clean transportation and battery systems company, proposed a battery tender that, when fully charged, would be able to provide 5 megawatt-hours (MWh) of electricity for traction purposes.[239] This amount of electricity is equal to the energy output of 4,400 THP locomotive operating at full power for 1.5 hours. A battery with this capacity should be large enough to provide some measure of hybrid capacity to diesel-electric freight locomotives operating on mountainous routes. It should also be suitable for regional passenger-rail applications that operate in stop-and-go fashion with rapid acceleration and deceleration between stations.

Battery tenders would work best in locomotive applications where recharging opportunities are frequent and short bursts of energy are needed. They could also supplement main-line freight hauling on certain routes. But for now, they will have to leave the bulk of demanding energy-intensive main-line freight operations to diesel fuel in part because of its high energy density. A full 5,000-gallon tank of diesel fuel contains 6.4 x 10^8 BTUs[240] or 188 megawatt-hours of energy—nearly 40 times the amount of energy envisioned for the energy tender described above.

Another drawback of battery tenders is their weight. Any tender with significant capacity would be heavy, which represents an energy penalty. Viewed alternately, the tender would supplant one revenue-producing car. They also must be coupled and uncoupled to locomotives (unless permanently attached).

Motorized Battery Tenders or ZEBLs

The term *Zero Emissions Booster Locomotive* (ZEBL)[241] has been coined to describe a battery tender with its own

electric traction motors, which produces no emissions of conventional air pollutants. This type of locomotive would meet California's definition of "zero emissions" because it does not emit NOx, hydrocarbons, particulate matter, and carbon monoxide at point of use. The illustration below shows a ZEBL coupled to a conventional diesel locomotive.

Once a ZEBL is added to the train, the train has been effectively hybridized. A ZEBL can store and provide regenerated braking energy while also delivering traction power to the rails through its own traction motors. Under certain circumstances, it might be able to provide effective electrification without the cost and visual impact of overhead wire catenary systems. Moreover, ZEBLs *could* provide low- or no-carbon (hereafter also referred to as *low/no-carbon*) performance

by running on electricity produced by regenerative braking and newly installed solar energy, wind power, or other low/no-carbon energy source.

According to ZEBL proponent, David Cook of Rail Propulsion, this type of motive power could produce significant energy savings, emissions reductions, and speed improvements at lower costs than purchasing new Tier 4 locomotives.[242] Cook envisions cab-equipped ZEBLs on the front or rear of regional rail trains, paired with cleaner burning diesel-electric locomotives whose fuel could be compressed natural gas to facilitate an eventual conversion to renewable hydrogen. Each ZEBL would have battery capacity equal to the energy content of 200 gallons of diesel fuel, begin service fully charged, and be recharged en route by regenerative braking and

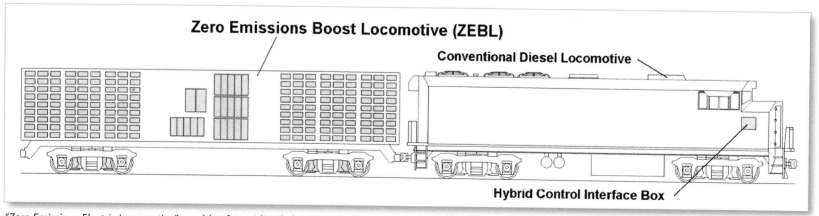

Zero Emissions Boost Locomotive (ZEBL)

Conventional Diesel Locomotive

Hybrid Control Interface Box

"Zero Emissions Electric Locomotive" capable of providing hybrid regenerative braking capability to conventional diesel locomotive. It also adds horsepower and tractive effort. Image Credit: Dave Cook.

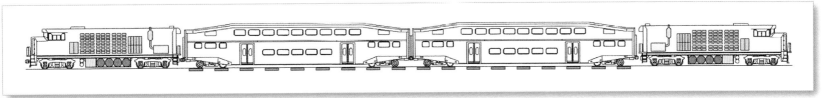

Regional commuter train with wireless power transfer receivers on all passenger cars. Power transmitters would be installed in the track roadbed the length of the station platform. These trains could operate with ZEBLs on both ends. Image Credit: Dave Cook.

wireless power transfer (WPT). To facilitate quick WPT "flash charging" during station stops, passenger cars would be equipped with multiple inductive power receivers that would be mounted on the bottom of passenger coaches stretching the length of the train. The receivers would be charged by inductive power transmitters located between track rails.

WPT is currently used for charging transit buses and electric cars, but it is still under development for rail applications by companies such as Bombardier, whose Primove electromagnetic induction system has been demonstrated on trams in Europe and China.[243] Primove trams also work off of catenary electric lines. Bombardier's WPT system uses continuous between-the-rail electrical conduit, segments of which are activated only when they are completely covered by a moving tram. Obviously, the amount of power wirelessly transferred to a tram is modest compared to what passenger and especially freight locomotives would require, but Bombardier has demonstrated the concept.

While CARB and RAILTEC estimate the cost of a battery tender with regenerative braking storage capability to be $3 million,[244] Cook sees lower costs especially when ZEBLs are remanufactured from older locomotives whose prime movers are removed to make room for batteries. Allowing buyers to lease ZEBL batteries would mitigate buyer concerns about rapid battery technology obsolescence.

6.4 Alternative Fuels for Diesel-Electric Locomotives

Locomotive manufacturers and railroads are now experimenting with alternative gaseous and liquid fuels that can be burned in locomotive diesel engines.[245] Actually, experimentation with alternative fuels has a long history in railroading, primarily motivated by a desire to find less expensive locomotive fuel. Alternative fuels under consideration here are:

- Natural gas
- Fischer-Tropsch Process fuels
- Liquid and gaseous biofuels

Natural Gas as Locomotive Fuel

Natural gas (methane, CH_4) can be used as locomotive fuel in two forms—compressed natural gas (CNG) or liquified natural gas (LNG). Natural gas locomotive fuel has been described as a path to reduced emissions and greater environmental responsibility—although, as we will see, the latter claim has been tarnished by methane emissions and hydraulic fracturing.

Despite discussion about natural gas within the railroad industry spanning decades, this alternative fuel has not achieved widespread use. Enthusiasm for it increases or decreases depending on the relative price of natural gas and diesel fuel. Other factors affecting interest in the use of natural gas as a locomotive fuel are the cost of providing new infrastructure and the delay in approval of safety standards for fuel tender designs.

While Class I railroads such as BNSF, Canadian National, and Union Pacific have tested LNG in diesel-electric locomotives, no Class I railroad is undergoing a transition to natural gas. On the other hand,

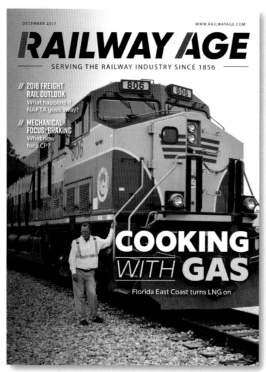

December 2017 cover of *Railway Age* magazine. Image Credit: *Railway Age*

Florida East Coast Railway's ES44C4 and GE Next-Fuel demonstrator separated by a LNG tender as they cross San Sebastian River, St. Augustine, FL. June 2016. Photo Credit: Michael Maiullari.

regional railroads Florida East Coast and Indiana Harbor Belt (IHB) are showing interest in using natural gas now.[246] In fact, Florida East Coast Railway has been retrofitting its ES44C4 fleet with General Electric LNG dual-fuel conversion kits and began LNG revenue operations in 2016.[247] (The railway's previous parent company owned a subsidiary that builds natural gas liquefaction plants.) At the same time, the IHB has been pursuing CNG as an alternative fuel for its switcher locomotives.

Here are some facts about using natural gas in diesel-electric locomotives:

- Tier 2 and 3 diesel-electric locomotives can be modified to burn LNG with GE's $400,000 - $500,000 Next Fuel™ retrofit kit, which has an 80% LNG/20% diesel fuel dual-fuel combustion capability.[248]
- Progress Rail/EMD is developing two dual-fuel LNG retrofit systems, Dynamic Gas Blending™ and High Pressure Direct Injected Gas™, which burn 60% and 95% natural gas blends, respectively.[249]
- LNG-fueled locomotives operating on 80% natural gas and 20% diesel fuel have been shown to significantly reduce NOx and PM compared to non-LNG-fueled Tier 2 and 3 locomotives, though they produce slightly greater amounts of these emissions than Tier 4 locomotives.[250]
- LNG use may reduce horsepower and efficiency compared to diesel fuel; a 10% decline in fuel efficiency is possible.[251]

- Burning 100% natural gas in a diesel engine requires spark ignition; blending natural gas with some fraction of diesel fuel provides ignition without a spark plug.
- LNG will not ignite as a liquid, and it boils off when spilled, but it does produce a violent reaction when combined with water.
- Compressing natural gas to fill a CNG tender can take hours compared to 25 minutes to fill a 10,000-gallon LNG tender, though competitive CNG refueling speeds are envisioned with new technology involving larger compressors and buffer storage tanks.
- Both CNG and LNG have additional energy requirements associated with the compression or liquefaction of natural gas—however, far more energy is required to chill natural gas to a liquid state than to compress it to useful pressures.
- While LNG has substantially more energy density than CNG, both have much less energy density than diesel fuel (see accompanying chart) and, as a result, require much larger fuel tanks than diesel fuel to carry the same amount of energy.[252]
- The liquefication process enables LNG to burn cleaner than CNG because it removes impurities when the gas is chilled.

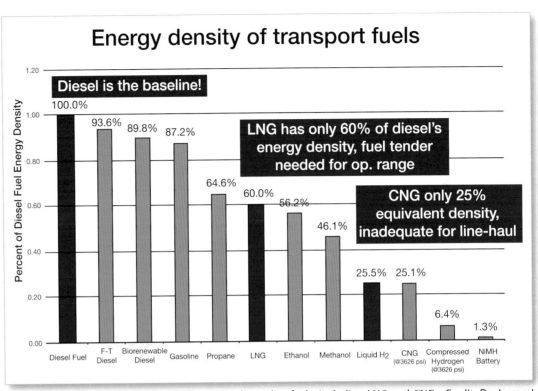

The high energy density of diesel fuel compared to other fuels, including LNG and CNG. Credit: Dr. James J. Eberhardt (US DOE).

- A CNG or LNG fuel tank could be installed in a tender attached to a locomotive or mounted between two locomotives at a cost of $800,000-$1.2 million, though costs would drop if a market for these tenders developed.[253]
- LNG tenders can hold 10,000 gallons in a tender similar to a tank car or 25,000 gallons in a more purpose-built tender capable of fueling a locomotive from Chicago to Los Angeles.[254]
- The payback of converting locomotives to LNG and providing them with shared LNG tenders is estimated to be 5 years,[255] assuming a natural gas price differential of $1.50 per "diesel gallon equivalent."[256]
- A switch to CNG or LNG would require costly infrastructure changes—e.g., refueling stations, pipelines, and compression or liquefaction stations that consume energy and reduce potential emissions benefits associated with natural gas as an alternative fuel.

- The cost of a natural gas liquefaction plant has been estimated to be $450 million for a million gallon per day capacity—enough to refuel 25 trains per day.[257]
- LNG tanks vent methane as they warm up if they are not connected to an operating diesel engine—thus, once the fuel tanks are filled, the LNG should be consumed.

But, importantly:

- While at the point of combustion, natural gas produces 25% less carbon dioxide per BTU than diesel fuel, and in locomotive engines may produce a 22.5% reduction of carbon dioxide (given a 10% reduction in engine efficiency), its use as a locomotive fuel does not reduce greenhouse gas emissions because

Leased Union Pacific LNG tender in Itasca Yard, Superior, WI, upon return by Canadian National after testing in Alberta, September 22, 2013. Photo Credit: David Schauer

of methane leakage in natural gas production, supply, and use.

Methane—A Powerful Greenhouse Gas.

Carbon dioxide and methane are both greenhouse gases that absorb heat and contribute to warming of the earth's atmosphere. But, comparing them on an equal mass basis, methane has 28-36 times the global warming potential of carbon dioxide over a 100-year time period, and, significantly, 84-87 times the global warming potential of CO_2 over the crucial 20-year time frame.[258]

The extreme global warming potential of methane would be of no consequence if all the natural gas produced was burned, and thus turned into carbon dioxide and water, but it is not. Some methane is released into the atmosphere when natural gas is produced, processed, and distributed. These emissions are called *upstream* or *Wells-To-Pump* methane emissions. There are also *downstream* or *Pump-To-Wheels* emissions for natural gas-powered vehicles. These emissions occur at refueling stations and during engine operation.

The EPA's current estimates of upstream methane leakage are in the 1.0% to 1.5% range. This sounds modest, but methane's global warming impact is significant for the reasons given above.

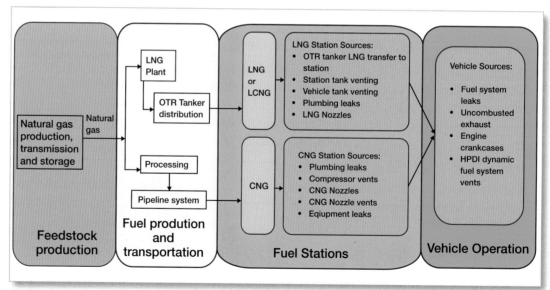

Methane emissions associated with natural gas-powered vehicles. The blue and yellow boxes show where upstream emissions occur, and the beige and green boxes show where downstream emissions occur. Methane slip occurs in the green vehicle operation box. Reprinted with permission from Environ. Sci. Technol. 2017, 51, 968-976. Copyright 2017 American Chemical Society.

However, there is disagreement over up-stream methane emissions. A 2014 study by Brandt et al. finds that official government estimates "consistently underestimate" actual methane emissions.[259] The Environmental Defense Fund's analysis of recent scientific studies of methane emissions has concluded that leakage rates are probably 1.5% to 2.5%.[260] Robert Howarth at Cornell University and colleagues have concluded that methane leakage associated with natural gas production is probably high enough to make the GHG footprint of natural gas greater than both diesel fuel and coal.[261]

The diesel-engine-based portion of downstream emissions is referred to as "methane slip." Slip occurs when unburned methane escapes from unvented crankcases (slipping around the piston rings) and exhaust stacks (because not all of the natural gas in engine cylinders is combusted). Newer diesel-electric locomotives have closed crankcases, so all methane emissions in these dual-fuel locomotives emanate from the locomotive's exhaust stack. Methane escapes combustion in engine cylinders as a result of:

- Valve overlap, which allows a small amount of the natural gas/air mixture to be exhausted as soon as it is injected into the cylinder
- Incomplete combustion in cylinder crevices, where flame quenching occurs at the cylinder walls

- Inadequate ignition temperature and gas-air mixture[262]

Methane slip in diesel-electric locomotive engines may be as high as 5% of the natural gas introduced into the engine, but it's reported to be between 1% and 2% with high-pressure direct injection (HPDI) LNG retrofit kits.[263] Engines designed to burn only natural gas can have less methane slip than HPDI.

At the levels of upstream and downstream emissions described above, the GHG benefit of natural gas locomotive fuel is essentially zero compared to diesel fuel. That said, it's worth noting that not all diesel fuel is equal when it comes to GHG emissions. If produced from tar sands, diesel fuel has a larger GHG footprint than when produced from conventional oil because of the additional energy required to convert this tarry substance into oil.

GE and EMD are aware of methane slip as it impacts natural gas use in diesel-electric locomotives. The manufacturers are said to be trying to reduce these emissions. However, while catalysts could oxidize at least some of the unburned methane in exhaust emissions, GE and EMD kits do not include them. The effectiveness of catalysts depends on exhaust temperature which is not always adequate throughout the locomotive's duty cycle. One study measuring

the effectiveness of a methane-reducing catalyst in heavy-duty diesel trucks found that at most they reduced methane emissions by 15%.[264]

Natural gas poses other environmental concerns. Over half of U.S. natural gas is called *shale gas* and is produced by hydraulic fracturing, an environmentally controversial drilling technique that involves pumping large quantities of water, sand, and chemicals under high pressure into very deep wells in shale rock formations to fracture the shale and release the gas.[265] Hydrofracking has these environmental impacts:

- Use of large volumes of fresh water
- Use of often unidentified hazardous chemicals
- Production of large volumes of toxic and radioactive wastewater
- Contamination of ground, surface, and drinking water when well casings fail or waste water is not properly treated and disposed of
- May be linked to increased frequency of earthquakes induced by water injection into deep wells, e.g., in Oklahoma[266]
- Industrialization of rural areas

These same impacts apply to shale oil, and thus to diesel fuel to the extent that diesel is produced from oil extracted by hydraulic fracturing.

Fischer-Tropsch Process Fuel as Locomotive Fuel

The F-T (Fischer-Tropsch) process works by gasifying coal, natural gas, or biomass feedstocks, and then liquefying constituents by a catalytic process. F-T synthetic diesel fuel has an energy density 7% less than conventional diesel fuel,[267] is low in sulfur, and has favorable auto-ignition attributes, which allow it to operate un-modified in diesel engines. Research on diesel trucks has shown that burning F-T fuel produces 12% less NOx emissions and 24% less particulate emissions than conventional diesel fuel.[268] However, even when produced with a biomass feed-stock, F-T process fuel tends to have a large carbon footprint because of all the energy that is used in the F-T fuel pro-duction process itself.

Biofuels as Locomotive Fuel

Biofuels are solid, liquid, and gaseous fuels derived from biological processes and sunlight, and, as such, are forms of renewable energy. The most likely liquid biofuels for use in diesel-electric loco-motives are "biodiesel" and "renewable diesel," both of which are produced by simple chemical reactions that trans-form vegetable and other non-petro-leum-based oils into organic liquid fuel that can easily be burned by a diesel en-gine. Significantly, these fuels potential-ly represent ways to run diesel-electric locomotives on solar energy. *Biogas,* also

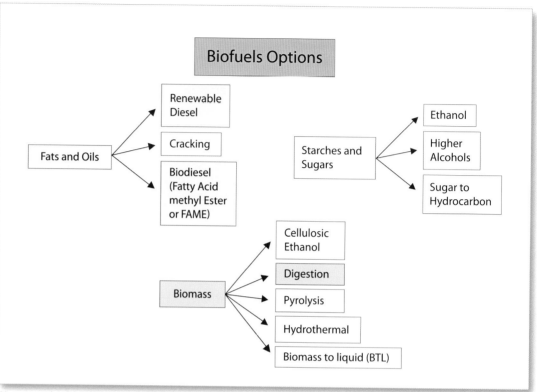

Numerous biofuel options and their sources. Biodiesel, renewable diesel, and renewable natural gas (depicted here as "digestion") are the most likely renewable biofuel replacements for diesel fuel to power diesel-electric locomotives. Image Credit: California Air Resources Board/Robert Hochberg.

called *renewable natural gas,* is a gaseous biofuel that could be used to power lo-comotives. It is created from anaerobic digestion of biomass.

Sustainable Biofuel. An acceptable, sus-tainable biomass-based fuel would:

- Burn efficiently and cleanly
- Be produced with minimal fossil fuels
- Produce minimal GHG emissions
- Not compete with food production

- Cause minimal damage to the en-vironment, natural resources, and human communities

Biodiesel. U.S. biodiesel is made from soybean oil, canola oil, distillers corn oil, waste cooking oils, and animal fats (a slaughterhouse by-product). Distillers corn oil is a by-product of the alcohol spirit and ethanol fuel industries. Biodiesel can also be produced through pyrolysis of cellulosic feedstocks (crops and forestry residues)

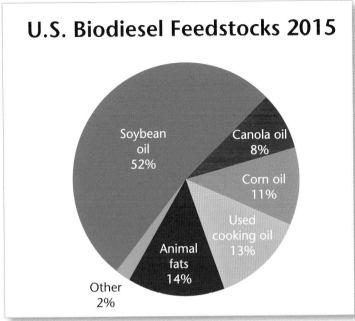

U.S. Biodiesel Feedstocks 2015

- Soybean oil 52%
- Canola oil 8%
- Corn oil 11%
- Used cooking oil 13%
- Animal fats 14%
- Other 2%

The percentages of various feedstocks used to produce biodiesel in the United States. Soybean oil is the primary feedstock. Image Credit: Union of Concerned Scientists/EIA.

and may eventually be made from algae grown in ponds or "bioreactors"—though this technology is not yet ready.

Biodiesel fuel production is closely tied to livestock production in at least two ways. As indicated above, animal fat from slaughterhouses is one feedstock for biodiesel. Also, the availability of soybean oil for biodiesel production is a function of the demand for soybeans for livestock feed. When soybeans are grown for livestock feed, the protein portion of the bean is used for animal feed while the "leftover" oil portion of the bean is available for biodiesel production. Thus,

under the current model, growth in biodiesel production would require growth in livestock production, which has been identified as a significant source of global GHG emissions.[269]

Biodiesel fuel is most commonly used in fuel blends such as B2, B5, B10, or B20, where the 2, 5, 10, and 20 describe the percentages of biodiesel in the fuel. The remainder is comprised of conventional petroleum-based diesel (referred to as *petrodiesel* in this section).

Biodiesel has been shown to provide engine responsiveness, horsepower, and fuel economy similar to petrodiesel in lower concentrations, e.g., B5 to B20. At higher concentrations, biodiesel performance declines somewhat in comparison.[270] The decline in power output at higher biodiesel blends is at least in part due to biodiesel's lower energy density when compared to petrodiesel. A gallon of biodiesel contains about 90% of the amount of energy in a gallon of petrodiesel.

Biodiesel locomotive fuel in all concentrations has been shown to reduce particulates, hydrocarbons, and carbon monox-

ide emissions compared to petrodiesel. But nitrogen oxides emissions have been a problem. Studies have shown that NOx emissions increase with fuel containing higher concentrations of biodiesel.[271]

Progress toward solving biodiesel's NOx problem may have been made on July 20, 2017, when the California Air Resources Board announced that it had certified a new biodiesel additive, VESTA™1000, capable of reducing B20 NOx emissions by 1.9% and particulate matter emissions by 18% compared to CARB certified petrodiesel.[272] More testing is needed to demonstrate applicability to locomotive diesel engines, but the National Biodiesel Board, which led the effort to produce this additive, says prospects look promising.

While railroads have been interested in biodiesel, the industry has been tentative about any kind of transition for various reasons including:

- Cost and availability
- Energy density
- Delivery, storage, and fuel infrastructure
- NOx emissions
- Gelling in cold temperatures
- Solvent effects at higher concentrations
- Need for durability testing
- Potential locomotive manufacturer warranty issues when used in newer locomotives
- Shorter shelf life than petrodiesel fuel

As explained in Section 2.4, diesel fuel gelling occurs when waxy constituents of the fuel begin to solidify as the fuel is subjected to colder temperatures. While biodiesel fuels made from different feedstocks "gel" at different temperatures, all the feedstock variants of biodiesel reach their gel points at temperatures higher than petrodiesel. That is a problem for cold weather operation when temperatures can easily fall below gelling points. Railroads can reduce gelling by specifying cold weather grades of biodiesel, using chemical additives, or reducing the concentration biodiesel in the fuel. B20 might be used in the summer, with a switch over to B5 or B2 in the fall in preparation for winter.

Biodiesel, like other biofuels, has the potential to be a low/no-carbon emissions fuel. "Carbon neutrality" would be achieved if the amount of carbon dioxide released into the atmosphere when biodiesel is produced and burned was equal to the amount of carbon dioxide absorbed by plants grown to produce biodiesel. Thus, the emissions associated with the production of biodiesel must also be taken into account. Currently, petroleum is used to power trucks and tractors, natural gas and coal to generate electricity used to crush soybeans, and natural gas to pretreat fats and cook the oil. These fossil fuel inputs contribute to biodiesel's carbon footprint but could be reduced.

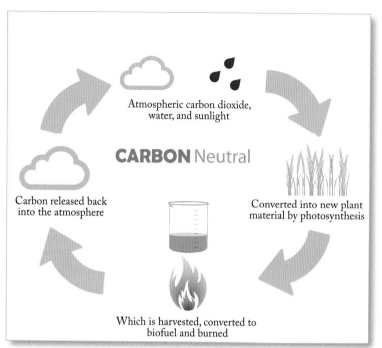

Theoretical "carbon neutral" biofuel carbon dioxide cycle. CO_2 is absorbed when plants grow for biofuel production and released when biofuel is burned. Image Credit: Robert Hochberg.

In 2010, the EPA reported that soybean, oil-based biodiesel reduced greenhouse gas emissions by 57% compared to a petrodiesel baseline.[273] Updated analyses indicate that the estimated reduction of GHG emissions may be closer to 75% or more.[274] The 2010 EPA report also concluded that biodiesel made from waste oils, fats, and greases reduces GHG emissions by 86% compared to petrodiesel. Of course, these percentages are based on a 100% biodiesel to 100% petrodiesel comparison. For reduced concentrations of biodiesel, the GHG benefit would be reduced accordingly. Thus, if B100

represents a 75% reduction in GHG emissions compared to petrodiesel, then B20 represents a 15% reduction.

According to the Union of Concerned Scientists (UCS), the demand for soybean oil associated with U.S. biodiesel production raises the price of soybean oil and indirectly leads to the expansion of production of other types of vegetable oil, here and elsewhere—including in nations like Malaysia and Indonesia where rainforests are cut down to create palm oil plantations.[275] Indirect land-use impacts like these, if they occur, would affect the carbon footprint of U.S. biodiesel. The National Biodiesel Board disagrees with the UCS analysis, arguing that because the soybean oil used for biodiesel in the United States is a by-product of the livestock feed industry, it does not affect vegetable oil pricing or production here or internationally.

In 2016, 2.9 billion gallons of biomass-based biodiesel were produced in the United States. There is disagreement about how much biomass production could increase. On the one hand, a UCS-commissioned study concluded that it would be reasonable to expect 29 million gallons per year of annual growth in U.S. biodiesel production from domestic sources.[276] On the other hand, the U.S. EPA and the National Biodiesel Board contend that the annual rate of growth could be over ten times that amount—up to 300 million gallons a year. That would represent a 10% growth rate.[277]

One potential obstacle to extensive biodiesel fuel use by the railroad industry is the reluctance of GE and EMD to maintain engine warranties (which cover materials and workmanship) when biodiesel is used to power their locomotives. Both manufacturers approve up to B5, but approval of higher blends is now subject to negotiation by each individual company. Through the Locomotive Maintenance Officers Association,[278] the National Biodiesel Board is working with locomotive manufacturers and railroads to share research and encourage approval of higher biodiesel blends. The NBB expects that higher blends will be approved soon by locomotive manufacturers.

Sustainably produced biodiesel can and should be an important alternative fuel for railroads. Only a fraction of eventual production, however, will be available for the railroad industry because of demand for it in other sectors of our economy—especially sectors like aviation that have fewer low-carbon alternatives. For that reason, and others, biodiesel is likely to provide only a partial solution to the railroad industry's quest for clean energy for its vast fleet of diesel-electric locomotives. Other, complementary, options will also need to be identified.

Renewable Diesel. While using similar feedstocks, renewable diesel and biodiesel are different fuels that are manufactured using different chemical processes.[279] Biodiesel is made from a process called "transesterification" of fats and oils, while renewable diesel is made from processes called *hydrogeneration* and *hydrotreating* of fats, oils, and esters. The Fischer-Tropsch process can also be used to make renewable biodiesel out of cellulosic biomass. Unlike biodiesel, renewable diesel is considered chemically equivalent to petrodiesel when it is produced in conformance with the petrodiesel fuel standard ASTM D975.

Renewable diesel is a relatively new fuel. Annual production volumes are much less and prices are higher than biodiesel. However, various advantages over biodiesel are claimed, including that renewable diesel:

- Is completely fungible (interchangeable) with petrodiesel and can use the same distribution system

- Can be more readily used in diesel engines in higher blends, i.e., RD100
- Contains negligible amounts of sulfur
- Has low cloud, gel, and pour points
- Has excellent long-term stability

The use of renewable diesel also produces emissions reduction of all criteria pollutants as well as carbon dioxide when compared to petrodiesel. According to CARB, if midwestern soybeans are used as the feedstock for renewable diesel, then only a 20% reduction in carbon dioxide can be expected. But CARB estimates that up to an 80% carbon dioxide emissions reduction can be achieved if the feedstock is slaughterhouse rendering—where the latter, while beneficial to recycle, is not a renewable energy source and raises the same issues discussed earlier about biodiesel, i.e., livestock production is itself highly carbon- and methane-intensive.[280]

Capitol Corridor trains running from San Jose through Sacramento to Auburn, California, are the first in the nation to run on renewable diesel.[281] The expected 67% reduction in GHG emissions suggests that the feedstock for this fuel will be primarily slaughterhouse-derived.[282]

Biogas, Renewable Natural Gas, or Biomethane. Landfills, cow manure, organic municipal solid waste, forest and agricultural residues, and wastewater treatment plants produce methane as a result of anaerobic decomposition. This methane is referred to as *biogas*, *renewable natural gas*, or *biomethane*, and should be captured and either flared (burned in the atmosphere) or used as fuel for power plants, vehicles, or other combustion devices. Flaring has a climate protection benefit because it converts the methane (which, as previously explained, has a high global warming potential) to less harmful carbon dioxide. Using captured biomethane as a fuel has a double GHG emissions benefit because it puts the methane to work (potentially displacing fossil fuel use) while converting it to carbon dioxide.

A recent study prepared for CARB concluded that with appropriate incentives California alone has the potential to produce 90.6 billion cubic feet per year of renewable natural gas.[283] This is enough energy to provide motive power to a significant number of locomotives operating in California, though its supply will be consumed by many other uses.[284]

6.5 Hydrogen Fuel Cells as an Alternative Prime Mover

While the diesel engine is a marvel of internal combustion power and efficiency, it may turn out that continuing its use in freight and passenger locomotives is not the best way to provide the industry with motive power while addressing climate change and other environmental concerns. If fuel cells were used as prime movers, then hydrogen—potentially produced from low/no-carbon emissions electrolytic, thermochemical, solar water splitting, or biological processes[285]—could be used as an alternative locomotive fuel. This could substantially reduce emissions of GHGs as well as conventional air pollutants—providing one path to cleaner locomotives and trains.

California Governor Arnold Schwarzenegger inspecting BNSF's prototype fuel cell locomotive with BNSF Chairman and CEO Matt Rose. This locomotive, built within the shell of a RailPower Green Goat, was powered by two 125 KW Ballard Power Systems fuel cells. The fuel cells plus a large on-board battery could deliver the equivalent of 2,000 hp. Photo credit: Bruce W. Jacobs.

With the addition of batteries, a fuel cell loco-motive could function as a hybrid locomotive with regenerative braking energy recovery. This kind of locomotive might initially be well suited for switching applications. Analogous fuel-cell-hybrid capability has been demonstrated in an underground mine loader that uses metal hydride hydrogen storage and a polymer electrolyte membrane fuel cell.[286]

How Fuel Cells Work

Fuel cells produce electricity by directly transforming chemical energy into electrical energy. They can be thought of as batteries that produce electricity as long as they are fed chemical fuel, typically hydrogen. Like batteries, fuel cells contain two electrodes, an anode and a cathode, which are catalysts where the chemical reaction takes place. The electrodes are separated by a material called the *electrolyte*, which permits positively charged hydrogen atoms to pass, and across which an electrical voltage is generated. When hydrogen and oxygen are supplied to a fuel cell, water vapor and heat are produced in addition to electricity. Thus, fuel cells can operate very cleanly.

A single fuel cell produces a very low voltage, e.g., less than 1 volt, so fuel cells are stacked and connected in series to produce the required voltage. Fuel cells are capable of very high energy efficiencies—60% to 70%, or more if the heat they produce is recovered and also put to use.

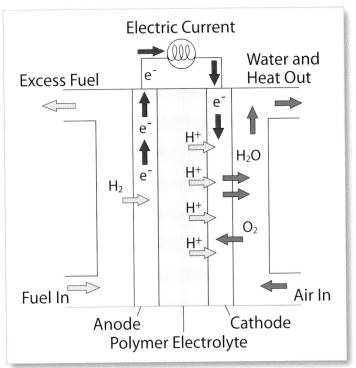

A polymer electrolyte membrane fuel cell schematic showing fuel, air, electric current, and exhaust flows. The fuel cell produces water vapor exhaust. Image Credit: Wikimedia Commons. (public domain)

The hydrogen consumed by fuel cells could be produced by "reforming" natural gas or coal,[287] gasifying biodiesel, "splitting" water into hydrogen and oxygen, and other means. But a fuel cell's hydrogen fuel would not be low/no-carbon emissions unless its source was low/no-carbon emissions. That rules out producing hydrogen from coal, oil, or natural gas unless carbon capture and sequestration (CCS) technology was used to produce electricity for hydrolysis. CCS, while technically possible, is energy intensive, expensive, and site specific.

But if electrolysis powered by wind- or solar-generated electricity was used to produce hydrogen fuel, then that fuel could theoretically be low/no-carbon emissions. Landfill, animal manure, and wastewater treatment facility biogas (also referred to as *renewable natural gas* or *bio-methane*) also represents potentially low-carbon sources of hydrogen fuel. Splitting high-temperature water into hydrogen and oxygen with solar concentrators or conceivably nuclear power represent possible future low/no-carbon emissions hydrogen fuel production methods.[288] Water splitting might also be accomplished using algae or semiconductors.

Hydrogen at atmospheric pressure has an extremely low energy density. Even in liquid form, its energy density is just 25% that of diesel fuel. Hence, a large fuel tender would be needed in order for a line-haul fuel cell locomotive to use hydrogen as a locomotive fuel. This tender would need to be superinsulated and pressurized to maintain hydrogen below its boiling point of minus 423°F.

According to the CARB Report, two types of fuel cells lend themselves to locomotive operation—the polymer electrolyte membrane fuel cell (PEMFC), and the solid oxide fuel cell (SOFC) with gas turbine (SOFC-GT).[289] Because of its lower energy density, CARB recommends the PEMFC for less demanding switching applications, while the more energy-dense SOFC looks more promising for line-haul locomotive applications. BNSF has been testing a prototype PEMFC switcher locomotive since 2009, reportedly with some success.[290] The PEMFC switcher was manufactured by retrofitting an old locomotive at an additional capital cost of approximately $3.5 million.[291]

The SOFC operates at a much higher temperature (1,200°F-1,800°F) than does the PEMFC. As such, it takes longer to warm up and produce meaningful power, but once it gets going, it has more power density, and when coupled to a gas turbine, can operate at a higher efficiency. The SOFC converts hydrogen fuel to electricity with 50% to 60% efficiency. Overall conversion efficiency can jump to 60% to 70% when the fuel cell is coupled to a gas turbine to recover high-temperature waste heat from the fuel cell. This type of fuel cell is referred to as a "SOFC-GT," and produces even higher effi-

ciencies if hydrogen is made on-board by reforming natural gas. While more efficient, a SOFC-GT using natural gas as primary fuel would produce a higher GHG footprint than a SOFC with a gas turbine running on renewably produced hydrogen.

Barriers to Fuel Cell Locomotives

The technical challenges facing fuel cell development for locomotive applications include:

- Energy density
- Power density
- Energy efficiency
- Weight
- Primary fuel source for the production of hydrogen
- Fuel delivery infrastructure
- Reliability in harsh railroad operating environment
- Long start-up time for SOFCs, preempting intermittent use
- Cost of fuel infrastructure
- Cost of fuel cell locomotives

CARB describes the SOFC-GT as in an "early research phase,"[292] and estimates that a locomotive-scale SOFC-GT would cost $6.5 million—with an additional $1.5 million needed to incorporate that power plant in a prototype locomotive.[293] Once in full-scale production, fuel cell locomotives would be expected to cost substantially less.

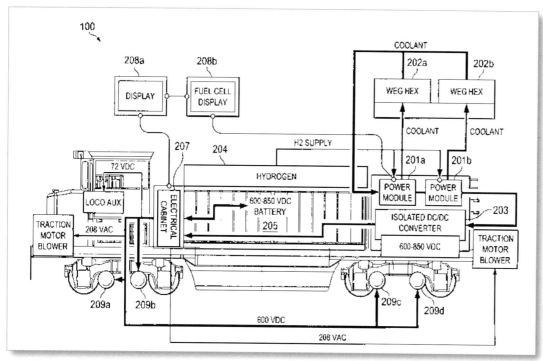

BNSF hydrogen fuel cell locomotive layout. Note hydrogen fuel tank, fuel cell power modules, and traction battery. Image Credit: BNSF patent (public domain).

Alstom's Coradia iLint hydrogen fuel cell passenger train ready for experimental service in Germany in 2018. Image Credit: ©Alstom/M. Wittwer

Alstom's hydrogen fuel cell-powered Coradia iLint passenger trains have two cars which can carry up to 300 passengers.[294] With a top speed of 87 mph, these potentially low/no-carbon "hydrail" trains will initially operate on a 60-mile route in northern Germany. Their fuel cell unit is mounted on the roof while lithium-ion batteries—storing excess fuel cell-produced electricity and energy recovered from regenerative braking—are located underneath the passenger compartment. A roof-mounted tank containing just 207 pounds of hydrogen is reported to be sufficient for 500 miles of operation. Hydrogen fuel for the test phase is being sourced from industrial emissions, though Alstom is said to be investigating more sustainable production of hydrogen. Alstom has obtained letters of intent to buy 60 more trainsets for use in Germany if prototype is successful.

While obstacles have stymied the development of fuel cell locomotives in the United States, the French multinational corporation that built France's high-speed TGV trains—Alstom—is overcoming those barriers and has announced that it is beginning service of the first hydrogen fuel cell-powered passenger train in 2018.

Of course, much more power would be required to operate freight trains, but the Alstom fuel cell passenger train represents an important development for this new technology.

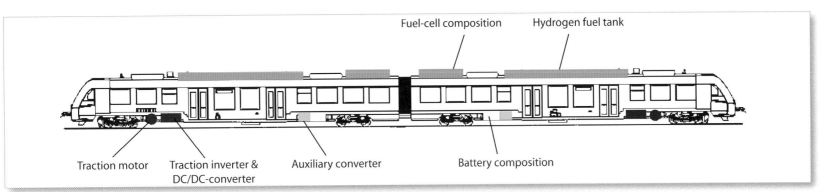

Locations of Alstom Coradia iLint fuel cell drive system components. Credit: @Alstom.

Ontario Metrolinx's Go Transit may take fuel cell technology one step further by using fuel cell locomotives as an alternative way to electrify six of its heavily used diesel-electric regional lines emanating from Toronto.[295] In February 2018, Metrolinx released its "Regional Express Rail Program Hydrail Feasibility Report" which concluded that (1) fuel cell motive power was technologically and economically feasible for Go Transit, and (2) the overall lifetime cost of building and operating a "Hydrail" system would be equal to that of conventional electrification.[296] While fuel cell locomotives would cost more than conventional electric locomotives, a Hydrail system would avoid the high capital costs associated with the installation of gantries and catenary wires, the construction of substations, raising bridges and lowering track bed to create catenary wire clearances, and the diversion of utilities.

Each Go Transit Hydrail locomotive would be capable of pulling 6 passenger coaches, with 2 locomotives required for 12-coach trains. These locomotives would be powered by a hybrid system consisting of a 2 MW PEMFC, 2 MW of battery capacity, and an ultracapacitor.[297] The fuel cell would charge the batteries and capacitor that would accelerate the train. The batteries and capacitor would also be charged by regenerative braking, potentially recovering more electricity than conventional regenerative braking. Hydrail technology would also be used to power electric multiple-unit cars.

The study recommends electrolysis to produce hydrogen for a Go Transit Hydrail system. The electricity for the hydrolysis process would be obtained from the regional power grid, consuming 1% of Ontario's electric supply—a small number representing a substantial amount of power. The report states that greenhouse gas emissions could be less than those of a conventionally electrified rail system because the hydrogen could be manufactured at night when electrical demand and fossil fuel generation in Ontario's power grid are low. In contrast, conventional electrified rail would consume electricity during peak daytime periods, when Ontario's fossil fuel power stations are running. Go Transit plans to announce its decision on Hydrail in 2019.

Other passenger railroads are showing interest in hydrogen fuel-cell-powered trains as a means of electrifying diesel-electric operation without the cost of catenary infrastructure. For example, in early 2018 Austria's Zillertal Railway published a request for proposals for this type of project.[298] The Zillertal Railway is a 20-mile long passenger line which runs through a picturesque valley. Here, catenary was regarded by some municipalities as unsightly.

6.6 Conventional Electrification

The term *conventional electrification* is used here to describe electrification using electric locomotives that receive power from overhead catenary wires. Thus, it is distinguished from electrification using battery tenders, fuel cells, or non-catenary means of transferring electricity to electric locomotives such as third-rail and wireless power transfer. Conventional electrification has many advantages and benefits but is considered here primarily because it provides a straightforward means of switching heavy haul railroad operation to low/no-carbon energy sources. To accomplish the latter, "decarbonized" electric power must be supplied to the power distribution systems serving electrified railroads.

How Electric Locomotives Work

Electric locomotives are simple in concept. They take electric power from overhead wires or third rails and use it to power traction motors that propel the locomotive and its train. Sophisticated electrical and electronic technologies enable these locomotives to operate with maximum power and energy efficiency. Like new diesel-electric locomotives, modern electric locomotives are equipped with AC traction motors.

Modern electric locomotives have the capacity for extensive regenerative braking. While dynamic braking used by conventional diesel-electric locomotives wastes the electricity generated by the locomotive's traction motors in braking mode, regenerative braking in electric locomotives captures and releases that electricity back into the locomotive's electrical distribution system—the overhead catenary—where it can be used by other locomotives. However, older electric locomotives also have dynamic braking grids, which burn off braking power when catenary line voltage is high and the lines are therefore "not receptive" to receiving electricity from regenerative braking. In contrast, Amtrak's newest electric locomotive, the Siemens 6,700 hp ACS-64,[299] does not have dynamic braking grids. It uses 100% of its regenerated braking energy either by returning it to the catenary or by using it on-board as head-end power (HEP).[300] Incidentally, Siemens reports that electric passenger locomotives are in traction motor braking mode 8% of operational time.[301]

Another approach for using 100% of regenerative braking energy is to use the catenary to collect and transfer it to temporary storage in wayside batteries or supercapacitors when there are no nearby locomotives or trainsets to use it.[302] This strategy—utilizing 30 wayside storage substations—is now being used by Philadelphia's SEPTA mass transit system. SEPTA (Southeastern Pennsylvania Transportation Authority) claims that just one of its substations receiving and storing energy in this fashion is saving $260,000 worth of electricity annually.[303]

Another recovery option is to sell excess recovered regenerative braking energy to the local power company.

Regenerative braking is just one feature contributing to the high energy efficiency of electric locomotives. Here is an estimate of the maximum design efficiency of current model electric locomotives, based on estimates of the maximum efficiencies of drivetrain components:

Main transformer	0.96
Rectifier	0.99
Inverter	0.99
Motors	0.96
Gearing and journal friction	0.99

Recovery from regenerative braking is >50% of kinetic energy available at time of braking.[304]

$$0.96 \times 0.99 \times 0.99 \times 0.96 \times 0.99 = 0.894$$
+ regen braking benefit

Given the benefit of regenerative braking energy recovery, it is reasonable to conclude that electric locomotive design efficiency is equal to or slightly more than 90%.[305] It has been reported to this author by an expert source to be "in the low 90s."[306]

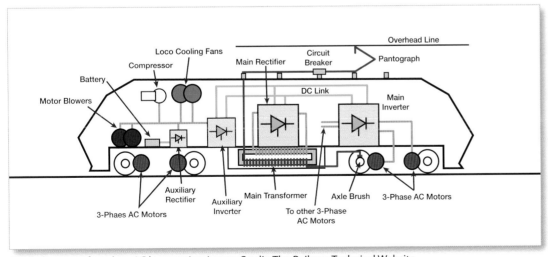

Block diagram of modern AC locomotive. Image Credit: The Railway-Technical Website.

Northbound *Acela* passes through Metuchen, NJ, station, March 2012. By 2022, Amtrak's 20 Bombardier-Alstom *Acela* trainsets are slated to be replaced by 28 Alstom *Avelia Liberty* trainsets. *Avelia Liberty* will operate on Amtrak's Northeast Corridor at speeds up to 160 mph (compared to the *Acela's* 150 mph) and use 15% less energy than the *Acela* primarily due to the *Avelia's* lighter weight. Photo Credit: Christopher Gore.

Thus, the efficiency of electric generation and supply is important. The efficiency of power generation is less an issue with renewable primary energy sources because, even where they are inefficient, no greenhouse gases are emitted except perhaps when wind turbines or solar panels are manufactured, shipped, and installed.

Cost/Benefit of Railroad Electrification

This outline summarizes the potential costs and benefits associated with converting diesel-electric railroads to electric locomotives and required infrastructure:

Costs

- Dollar, environmental, and other costs associated with acquisition of:
  - Electric locomotives
  - Catenary and substation systems
  - Power distribution systems
  - Additional low/no-carbon energy capacity
- Dollar costs associated with creation of sufficient overhead clearance for catenary, e.g., in tunnels and under bridges (in addition to those clearances already provided for double-stack trains)
- Loss of free interchange of locomotives throughout railroad network

Benefits

- Potential to operate on low/no-carbon emissions energy sources
- Improved energy efficiency with use of regenerative braking

When evaluating the efficiency of electric locomotives, it is also important to consider how much energy is used and wasted producing and distributing the electricity they consume. If the electricity supplying the catenary is produced by a thermal generating station, then as much as *two-thirds* of the heat produced by the primary fuel—coal, oil, natural gas, or nuclear—may be lost in the power plant's conversion of primary fuel energy to electricity (though some natural gas combined cycle power plants are now reported to be operating at greater than 60% efficiency or with less than 40% waste).[307] Additional energy is lost in utility transmission and railroad distribution systems, each of which may be 90% efficient.[308] When all losses are included, the overall worst-case efficiency of an electric locomotive could be less than 30%:

$$0.33 \times 0.90 \times 0.90 \times (0.96 \times 0.99 \times 0.99 \times 0.96 \times 0.99) = 0.24 \text{ or } 24\%$$

- High tractive effort and horsepower
- Faster acceleration
- No loss of power in tunnels
- Need for fewer locomotives
- Reduced locomotive maintenance
- Quieter operation
- Lower fuel costs
- Lower emissions overall, with zero emissions at point of use

Electrification of Freight Railroads

We see railroad electrification in the United States today on the passenger side—in subway, streetcar, and light rail systems, some regional commuter rail lines, and Amtrak's Northeast Corridor (NEC) which runs 450 miles between Boston, Massachusetts, and Washington, D.C., and accommodates 125-150 mph electric passenger train service. But to what extent could railroad electrification be expanded? Specifically, in light of the climate crisis, what are the eventual prospects for electrifying U.S. freight railroad operations?

Norfolk & Western coal train pulled by a 3,200 hp LC-1 electric boxcab locomotive, Switchback, WV. April 1929. Photo Credit: NWHS James N. Gillum Archives Collection.

There are numerous early examples of U.S. electrified freight railroads. Famous among them are the electrified routes of the Milwaukee Road, the Great Northern Railway, the Virginian Railway, the Norfolk & Western Railway, and the Pennsylvania Railroad. Among the reasons these railroads undertook electrification were to (a) reduce pollution in populated areas, long tunnels, and terminal areas, (b) increase traffic density, and/or (c) provide improved adhesion to pull heavy trains through mountainous terrains. While these early experiments in freight electrification did not last long once dieselization had occurred, they still proved the value of electrification.[309]

The idea of electrifying U.S. freight railroads re-emerged during and after the energy crises of 1973 and 1979 as a means of reducing foreign oil dependence and bolstering national security.[310] At that time, diesel fuel would have been replaced by coal to generate electricity. However, reconsideration of electrification never went anywhere because of the high cost of electrifying and the unresolved question of "Who will pay?"[311]

A conceptual rendering prepared for Virginia Rail Solution that envisions high-speed freight and passenger electric rail near Roanoke, VA. Copyright J. Craig Thorpe, used with permission of the artist.

Modern electric operation of freight railroads is the vast exception in the United States,[312] but it is well established in Russia and some European nations. In fact, Germany is presently considering additional electrification of freight railroad lines in support of its climate protection goals,[313] and the Italian State Railways freight subsidiary Mercitalia will be operating high-speed electric locomotive-powered freight trains beginning in October 2018.[314] But the conversion of U.S. freight railroads to electric motive power has seemed like

an impossible task. The costs of converting Class I railroads' over 160,000 miles of track (93,000 miles of "road"[315]) and 26,000 locomotives[316] to electric operation would be staggering. A more realistic proposal would focus on a portion of railroad corridors that are most heavily used.

Suppose, for the sake of argument, that 20% of U.S. freight railroad route-miles carried 70% to 80% of U.S. freight tonnage (revenue ton-miles). Or that 5% of the route-miles carried 25% of freight

tonnage. These scenarios can provide good starting points for discussing and evaluating U.S. freight railroad electrification. As route-mile percentages increase, it becomes a matter of diminishing returns (tonnage vs. cost), and, thus, more difficult to financially justify the infrastructure electrifications costs.

The electrification of U.S. freight railroads has most recently been explored by *Solutionary Rail*,[317] a new book that calls for electrifying 4,400 track-miles of BNSF's Chicago to Seattle northern transcontinental route. The authors—citizen activists, railroad union members, indigenous peoples, and railroad experts—envision clean, energy efficient, high-speed freight operation powered by renewable energy. The electric infrastructure for this new rail system would be funded by public/private partnerships and would be publicly owned and operated. Additionally, electrified rail corridors would "co-deploy" or share space with power transmission lines to facilitate the development of renewable energy in remote or grid-constrained parts of the country.

Solutionary Rail is an outgrowth of a similar proposal by Rail Solutions and the Steel Interstate Coalition that called for the electrification of the eastern rail corridor running adjacent to Interstate 81 in Virginia and other states.[318] Another interesting proposal by rail analyst Brian Yanity examines the costs, benefits, and feasibility of electrifying freight railroading in Southern California[319] These proposals point in the right direction but need serious consideration by the railroad industry and government.

Estimating the Cost of Partial Electrification

Here is a rough order-of-magnitude estimate of the capital cost to electrify 5% of U.S. freight route-miles in order to capture (according to our hypothetical scenario) 25% of U.S. freight tonnage.

Electric Locomotives. If a new electric locomotive costs $5 million,[320] then replacing enough locomotives to handle, say, 25% of U.S. rail freight could be estimated as follows:

26,000 locomotives x 0.25 x 0.8 = 5,200 new locomotives @ $5,000,000 = $26 billion

The 0.8 factor above reflects the greater availability and tractive advantage of electric locomotives compared to diesel-electrics, and, therefore, the need for fewer electric locomotives to perform a given amount of work.

This high locomotive replacement cost must be viewed in the context of providing the railroads with all new locomotives with long operational lives ahead of them, reduced maintenance requirements, potentially lower fuel costs, and clean point-of-use operation.

Electric Power Distribution System. If the cost of installing electric catenary is $4.1

Past and present Amtrak electric locomotives. *On the left* AEM-7 #905 pulls northbound Amtrak Northeast Regional into the Wilmington, DE, station, April 2014. These 7,000 hp locomotives were manufactured by General Motors between 1978 and 1988. Defying aerodynamics, the box-shaped AEM-7s were easily able to pull passenger trains 125 mph on the Northeast Corridor. By 2016, Amtrak had replaced all its AEM-7s with new 6,700 hp Siemens ACS-64 locomotives, shown *on the right* pulling another northbound Northeast Regional into the same train station in July 2015. Photos by author.

million per route-mile,[321] then installing catenary on 5% of Class I railroad's 93,000 route-miles could be estimated as follows:

$$\$4.1 \text{ million per mile} \times 0.05 \times 93,000$$
$$\text{miles} = \$19.065 \text{ billion}$$

Total Cost. This yields an estimated total cost of $45.065 billion ($26 billion + $19.065 billion) to electrify 5% of U.S. Class I railroad route-miles in order to haul 25% of Class I freight tonnage. Note that power distribution costs could be much higher or lower than the estimated $19.065 billion, depending on various factors. For example, costs might be much higher if routes were severely impacted by low clearance at overpasses, in tunnels, etc.—a likelihood for railroads operating in more populated or mountainous settings.[322] Also, this estimate does not include the disproportionate costs per track-mile to electrify sidings and yards.

If this were a rigorous calculation of costs, then life cycle locomotive maintenance costs and longevity would also have to be quantified and factored into the calculation. While electric locomotives require significantly less maintenance than diesel-electric locomotives, some of that edge is lost by the high cost of parts and catenary maintenance.[323] While both diesel-electric and electric locomotives have long lifespans, diesel-

electric locomotives might be hard pressed to function for as many miles as electric locomotives—for example, Amtrak's venerable fleet of AEM-7s traveled 3 to 4 million miles each in more than three decades of revenue service. Of course, freight duty is probably more difficult on a per-mile basis.

A $45 billion price tag gives pause, but it really would be a relatively small price to pay to convert 25% of U.S. rail freight to electrified operation, and, as such, to potentially low/no-carbon emissions operation. But, again, who pays? Presumably, costs associated with electrifying freight railroads would be paid by some combination of public and private funding, including perhaps from federal government revenues generated by a national carbon tax or fee,[324] or by a similar climate protection mechanism.

In addition to electrifying a subset of the freight lines, there are other approaches to partial electrification. Partial electric operation could also involve these elements:

- Use of dual-mode locomotives capable of either catenary or diesel/biodiesel operation[325]
- Integration of high-capacity battery tenders or battery locomotives that would:
  - Permit electric operation with no catenary or with intermittent catenary
  - Allow hybrid/regenerative braking operation with no catenary
  - Be pre-charged in terminals or from catenary with new wind or solar power

Dual-mode passenger locomotives are already in use. These include Amtrak's GE Genesis P32AC-DM locomotives and the Long Island Railroad's EMD DM30AC locomotives. Both of these are diesel-powered but function as electric locomotives powered by third rail in New York City rail tunnels and stations, and in the case of the P32AC-DM, from Croton Harmon, New York, to NYC's Penn Station. Amtrak may purchase new dual-mode locomotives capable of running on the Northeast Corridor as electric locomotives and then switching to diesel-electric mode after leaving the NEC in Washington, D.C. This would eliminate the need to switch locomotives at that point.

Another dual-mode locomotive is New Jersey Transit's Bombardier ALP-45DP, which is diesel-powered on non-electrified NJT routes but functions under catenary as an electric locomotive on NJT electrified routes and when in NYC rail tunnels and stations.[326] The ALP-45DP has enough power in electric-mode (5,360 hp) to serve as a freight locomotive but insufficient tractive effort and adhesion. However, very capable dual-mode freight locomotives could be designed and produced—likely with 3-axle trucks and additional weight.

A Bombardier dual-mode ALP-45DP, running under catenary. The powerful locomotive is seen here pulling a westbound NJT commuter train past North Elizabeth Station on Amtrak's Northeast Corridor. Photo Credit: Christopher Gore.

How Many Wind Turbines Would It Take?

Under our 5%/25% scenario, how many utility-scale wind turbines would be needed to displace the diesel fuel that diesel-electric locomotives now use to move 25% of U.S. Class I railroad freight tonnage? Here is a very rough estimate:

Assumptions

- Class I railroads annually consume 3.7 billion gallons of diesel fuel.[327]
- Hauling 25% of freight tonnage requires consuming 25% of the above fuel volume.

- 1 gallon of diesel fuel contains 128,000 BTUs of energy (LHV).
- Diesel-electric locomotives annual average operating efficiency is 33%.
- Revenue ton-miles of freight consume the same amount of energy regardless of what route they are on.

- 1 kilowatt hour of electricity equals 3,412 BTUs of energy.
- 1 megawatt-hour equals 1,000 kilowatt hours.
- Electric locomotive annual average operating efficiency is 85%.
- Utility electric transmission system is 90% efficient.[328]
- Railroad electric distribution system is 90% efficient.[329]

- Each wind turbine produces 3.6 megawatts at full power.
- Wind turbine availability is 33% ("availability" here means the percentage of full power capacity that the turbines generate annually).
- Number of hours per year is 8,760

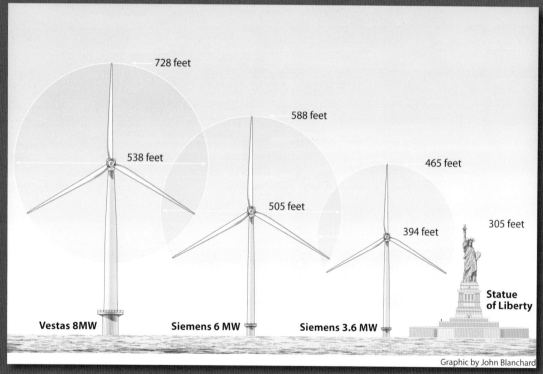

538 feet
728 feet
588 feet
505 feet
465 feet
394 feet
305 feet

Statue of Liberty

Vestas 8MW
Siemens 6 MW
Siemens 3.6 MW

Graphic by John Blanchard

The size of large wind turbines compared to the height of the Statue of Liberty. The turbines assumed in the wind turbine calculation are the 3.6 MW turbines, whose rotor hub is about the same height as Lady Liberty's crown and whose blades reach 160 feet higher. Image Credit: John Blanchard, originally published in Sierra Magazine, 2015 .

Annual amount of work, in BTUs, performed by diesel-electric locomotives consuming 25% of 3.7 billion gallons of diesel fuel:

3.7 billion gallons/yr x 0.25 x (128,000s BTUs/gallon) x 0.33 eff. = 39.072×10^{12} BTUs/yr

Annual amount of work performed by diesel-electric locomotives in MWhs:

$$\frac{39.072 \times 10^{12} \text{ BTUs/yr}}{3,412 \text{ BTUs/kWh} \times 1,000 \text{ kWh/MWh}} = 11,451,348 \text{ MWh/yr}$$

Amount of electricity wind turbines must provide to the utility transmission system annually to power electric locomotives to produce 11,451,348 MWh of work/yr:

$$\frac{11,451,348 \text{ MWh/yr}}{0.85 \text{ eff.} \times 0.90 \text{ eff.} \times 0.90 \text{ eff.}} = 16,632,314 \text{ MWh/yr}$$

Number of 3.6 MW turbines required to generate 16,632,314 MWh/yr:

$$\frac{16,632,314 \text{ MWh/yr}}{8,760 \text{ hrs/yr} \times 3.6 \text{ MW} \times 0.33 \text{ availability}} = \textbf{1,598 turbines}$$

Not surprisingly, this is a large number of giant utility-scale wind turbines, but they are easier to imagine as 16 wind farms, each with 100 turbines and located in different parts of the country. They could easily be accommodated in the Great Plains of the Midwest and offshore. If larger wind turbines were used in this calculation, far fewer turbines would be needed, but going bigger makes siting them more difficult.

Low-or-No-Carbon Power for Electrification

Electricity that is produced from renewable resources is often called *green power*. Electrified railroads could *self-generate* (build their own renewable electricity generation) or buy *green power*.

One of the most interesting examples of railroad self-generation is the solar tunnel running alongside a highway in Antwerp, Belgium. The roof of this tunnel is covered with 16,000 photovoltaic solar panels, stretching 2.2 miles and generating 3,000 megawatt-hours of electricity a year. This was reported to be enough electricity to power all of Belgium's trains for . . . *a day*.[330] Yes, for just one day—an astonishing statistic given the massive size of this solar array. Three hundred and sixty-five similar installations would be required to generate enough solar power to operate Belgium's trains annually.

Another example of railroad solar self-generation is the large solar installation serving Blackfriars subway station, in London, England. The station is located on a bridge spanning the Thames River. This impressive and architecturally beautiful photovoltaic array covers the entire train station and bridge and produces enough electricity to provide 50% of the power used by . . . *the station*.[331]

Buying Renewable Electricity. The Belgium and U.K. examples make clear the challenge of running electrified rail on self-generated solar power or solar power generally. Wind power has a greater potential to supply large amounts of renewable electricity, but its scale makes it less likely to be considered by railroads for self-generation. Given these realities, buying renewable electricity might seem like a better fit for railroads interested in powering electrified rail systems sustainably.

Renewable electricity can be purchased by buying "Renewable Energy Certificates," or "RECs," on what is called the *voluntary market*. RECs are market-based products originally designed to track and encourage renewable energy production. Technically speaking, they are tradable instruments that convey ownership of *the claim* that 1 megawatt-hour of renewable electricity was generated and delivered to an end-user or the grid.[332] Less technically, RECs have been described as the "environmental attributes" of (or, less formally, the "bragging rights" for) 1 MWh of renewable electricity generated and delivered to an end-user or the grid.

Pre-construction illustration of Blackfriars station with 4,400 solar panels. Solarcentury completed the installation in 2014, making it the world's largest solar bridge. Photo Credit: Network Rail.

RECs can be a little hard to grasp! For example, a REC is *not* the same thing as the megawatt-hour of electricity associated with it. When a REC is purchased, the generator still owns the underlying electricity unless the REC buyer also contracts to purchase that energy.

Buying only RECs is unlikely to produce an environmental benefit. The underlying electricity should be purchased as well—preferably using a long-term power purchase agreement (*PPA*) tied to the construction of new additional renewable capacity. Also, if the buyer's goal is—as it should be—to reduce GHG emissions, it is desirable to buy renewable electricity in a carbon-intensive regional grid (with, for example, a lot of coal-fired electricity) because when a renewable generator sells a MWh of electricity from renewable sources into this kind of grid, it is more likely to displace more carbon-intensive generation and greenhouse gases than if that MWh was sold into a less carbon-intensive grid.

Perhaps the best way for a railroad to buy renewable electricity would be for it to work directly with a developer who would build new renewable energy capacity specifically to serve the railroad. The railroad could lease some of its property to a wind or solar developer and share the financial benefits of a long-term PPA with the developer.

Now let's look at two instructive examples of railroad green power purchasing—one under consideration in Canada, and the other pledged in Germany.

Canadian VIA Rail is considering a plan to electrify a 600-mile route between Montreal, Quebec, and Windsor, Ontario.[333] However, the climate protection claims of this project could be called into question if VIA Rail's new electric trains were powered by REC purchases, or even power purchases, from *existing* hydroelectric facilities.[334] Why? Because buying existing renewable energy capacity may only reassign that capacity from one customer set to another, and that does not reduce fossil fuel burning or GHG emissions. In order to produce a GHG emissions benefit, VIA Rail's project must shift the regional power generation fuel mix away from fossil fuels. The only way it can do that is by causing *new* renewable energy capacity to be developed—hopefully in equal proportion to the electric load of its electrification project.

Moreover, if the electricity for this project was sourced from Hydro-Quebec's northern "James Bay Project," it is worth

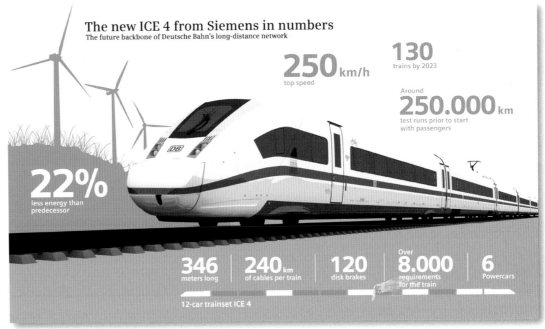

The new ICE 4 from Siemens in numbers
The future backbone of Deutsche Bahn's long-distance network

250 km/h
top speed

130
trains by 2023

Around
250.000 km
test runs prior to start with passengers

22%
less energy than predecessor

346
meters long

240 km
of cables per train

120
disk brakes

Over
8.000
requirements for the train

6
Powercars

12-car trainset ICE 4

German railway Deutsche Bahn has made commitments to purchasing green power to offset its greenhouse gas emissions. Image Credit: Siemens Corporation.

remembering that this project was highly controversial when it was being constructed decades ago.[335] Rivers were redirected, vast forested land areas were submerged by reservoirs, caribou migration routes were severed, and Cree and Inuit native peoples lost homelands and livelihoods. Plus, once under water, decomposing forests released methane and carbon dioxide—a climate-change-fueling effect not generally associated with renewable electricity generation.

Deutsche Bahn (DB), Germany's heavily electrified passenger and freight railway, states that it's strongly committed to purchasing green power—to offset the greenhouse gas emissions associated with its annual consumption of 12 terawatt-hours (12 million MWh) of electricity.[336] Previously criticized for its reliance on nuclear and then coal power, DB switched gears. Corporate policy now calls for a 30% reduction of GHG emissions by 2020, compared to 2006 levels, and a 100% reduction by 2050. In order to accomplish these goals, DB must ensure that its green power purchases produce new renewable capacity, and that the renewable electricity generated by this capacity offsets the GHG emissions of fossil fuel generation.

Providing enough renewable electricity to electrify railroads, especially those hauling freight, is a challenge because these railroads consume a lot of energy to move long, heavy trains over all kinds of terrain.

Consider that a single 4,400 THP locomotive can deliver nearly 4,000 horsepower to the rails, and freight trains typically have multiple locomotives pulling and pushing their loads. Four thousand horsepower is roughly equal to 3 megawatts, which, in turn, equals the full peak power output of a giant wind turbine or 150,000 square feet of solar panels—not considering the intermittency and variability of wind and solar. Diesel fuel is energy-dense while wind and solar energy are not.

A Role for Nuclear Power? While controversial, new nuclear power plants could in theory provide carbon-free electricity for railroad electrification. In this case, railroads might use a long-term PPA to directly purchase electricity from the operators of new nuclear power plants or become a nuclear power customer by purchasing it off the grid. Let's consider the current debate, and, thus, the appropriateness of building new nuclear power plants to meet the kind of large new electric loads that would be created by additional railroad electrification.

Climatologist James Hansen leads a group of scientists who argue that a new, safer generation of nuclear power plants can and should be built, and are essential to providing a credible path to climate stabilization.[337] In their view, the climate change threat is just too great to take potentially viable options off the table.

While fully supporting renewable energy, Hansen and his colleagues don't believe solar, wind, and other renewable energy technologies will be adequate to meet all of our electrical energy needs in the foreseeable future even if we got more serious about energy efficiency and conservation.[338] Hansen argues that nuclear power is needed to provide low/no-carbon baseload electrical power to meet current electrical demand and also to meet anticipated new electrical demand associated with the forthcoming transition to electric vehicles and the needs of developing countries, the world's poor, and a growing human population. Hansen's view finds support in U.S. Energy Information Administration data, which reveal that in 2016 only 15% of U.S. electricity was produced by renewable energy sources, with half of that from hydro, even after years of promoting electricity from solar and wind.[339] Switching to renewable energy for building space heating and rail freight and passenger transportation is even more challenging.

Supporters of nuclear power point to the fact that no other energy source comes close to the 57% share of U.S. carbon-free electricity now provided by nuclear power.[340] They also point to facts such as this: One large central nuclear power station can produce the same amount of electricity as 1,000 huge 3 MW wind turbines and can do so without the intermittency of wind or other renewables.

Will atomic energy power tomorrow's railroads?

Some day you may see a train like this — powered by the energy locked up in the atom.

Possibly the locomotive will have its own nuclear reactor. Or perhaps it will use electricity generated at atomic power stations. But this much is certain. Of all forms of land transportation, railroads offer the greatest opportunities for the efficient use of nuclear energy.

Railroads are constantly exploring exciting possibilities like this. Such progressive thinking is important to all of us — for we're going to need railroads more than ever in the boom years ahead.

Clearly, it's in the national interest to give railroads equal opportunity and treatment with other forms of transportation. America's railroads — the lifeline of the nation — are the main line to *your* future.

ASSOCIATION OF
AMERICAN RAILROADS
WASHINGTON 6, D. C.

This 1948 advertisement suggesting that nuclear power could be used to power trains—either with on-board nuclear reactors or electricity generated by "atomic power stations." Image Credit: Association of American Railroads.

The United States has 100 operational nuclear power plants, most built in the 1970s and 1980s. These are considered "Generation II" nuclear power plants—with earlier prototype plants categorized as "Generation I." Hansen would like to see improved "Gen III" light water nuclear plants built now while R&D on "Gen IV" nuclear power plants is accelerated.[341] While there are a number of different Gen IV designs under consideration, all would have these design features:

- Automatic emergency shutdown
- Passive reactor core cooling in the event of a shutdown
- Fueled by nuclear waste from both Gen II and Gen III nuclear power plants and nuclear weapons production/dismantling programs
- 100 times as efficient[342] as Gen II and Gen III plants—which would make fuel supplies of existing nuclear waste sufficient to last for nearly a thousand years
- Minimal additional nuclear waste generation; and the waste that is produced would have a much shorter half-life
- Closed fuel cycle system, less prone to diversion and nuclear weapons proliferation

Hansen believes that since the 1990s we have had the expertise to build a Gen IV power plant. He would like to see that project finally move forward, and has called for a nuclear renaissance.

Meanwhile, other scientists and environmental organizations remain steadfastly opposed to nuclear power. Amory Lovins of Rocky Mountain Institute has long pointed to problems throughout the nuclear fuel cycle (e.g., safety, fuel enrichment, waste disposal, diversion, weapons proliferation, etc.) but says that the high

Generation IV: nuclear energy systems deployable no later than 2030 and offering significant advances in sustainablity, safety and reliability, and economics.

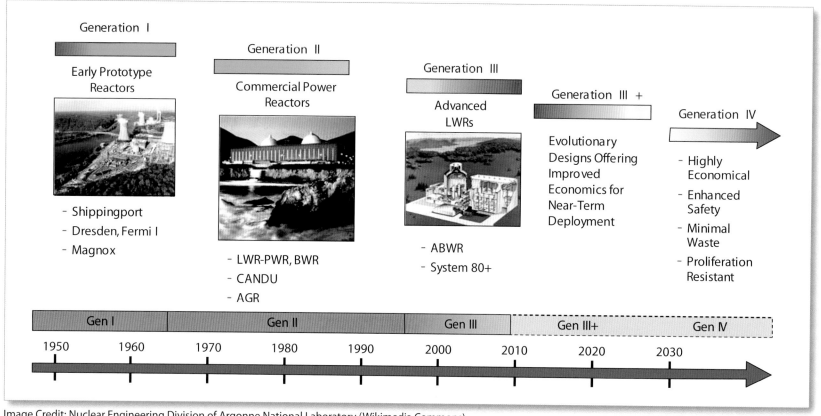

Image Credit: Nuclear Engineering Division of Argonne National Laboratory (Wikimedia Commons).

cost of nuclear power compared to energy conservation and renewable energy sources is alone enough to disqualify it.[343] Joe Romm of Climate Progress,[344] Stanford University professor Mark Jacobson,[345] and others have made the same argument. Jacobson, who has developed plans to meet 100% of U.S. energy demand with renewable energy, also argues that nuclear power plants are actually

fossil-fuel-intensive (given the amount of fossil fuel consumed in the construction of nuclear plants and production of nuclear fuel), not needed for grid stability, and would take too long to design, site, and build to make a difference soon enough for climate protection.

In a way, the U.S. nuclear power industry never recovered from the 1979 nuclear

accident at Three Mile Island, near Harrisburg, Pennsylvania. The Chernobyl accident in 1986 and the Fukushima, Japan, accident in 2011 further undermined public attitudes toward nuclear power, contributing to a long-standing stalemate about building new U.S. nuclear plants.[346] In 2018, concerns about nuclear power were reinvigorated by revelations that the computer control systems of U.S. nuclear power plants had suffered a cyberattack by Russia, giving the United States' principal adversary the ability to sabotage or shut down these plants at will.[347]

6.7 Carbon Credits to Offset Carbon Emissions

The creation or purchase of carbon credits represents another strategy railroads could use to address GHG emissions from their locomotive fleet—irrespective of motive power type. Carbon credits are often called *offsets* because they offset or have the effect of balancing or canceling out GHG emissions that are occurring elsewhere.

Technically speaking, a carbon credit is a market-based sellable or tradable certificate representing the reduction, avoidance, or sequestering of 1 metric ton (2,200 pounds) of carbon dioxide or its greenhouse gas global-warming-potential-equivalent (tCO$_2$e). The concept of equivalence is used to allow for credits comprised of GHGs other than carbon dioxide. For example, assuming methane has a global warming potential 28 times that of carbon dioxide (see discussion of methane in Section 6.4), then a carbon credit comprised of methane would represent 2,200/28 or 78.6 pounds of methane reduction, avoidance, or sequestering.

The official process for buying, selling, and trading carbon credits at the national level was established by the 1997 Kyoto Protocol to the United Nations Framework Convention on Climate Change. Kyoto established legally binding commitments for developed countries to reduce greenhouse gases in order to address climate change. The United States signed but did not ratify this agreement.

The purchase of carbon credits typically takes place within a "cap and trade" regime that permits GHG emitters to buy carbon credits in order to stay below their emissions caps. A voluntary market arose alongside Kyoto's compliance markets, including in the United States. The voluntary market exists so that businesses (including railroads), non-governmental organizations (NGOs), and individuals can voluntarily buy credits to offset their own emissions.[348]

Among other criteria, a valid carbon credit should be:

Real—Tied to an actual project with third-party monitoring, reporting, and verification

Enforceable—Protected by a mechanism that would penalize the credit producer if the project is not completed or managed as promised

Additional—Producing GHG emissions reductions *in addition* to what would have otherwise occurred in a "business-as-usual" scenario[349]

Purchasing carbon credits can be an effective way to neutralize portions of a carbon footprint if care is taken to ensure the validity of the credits and the projects producing them. Ideally, carbon credit purchasing would be preceded or accompanied by on-site emissions reduction efforts.

Railroads interested in offsetting their carbon emissions could also develop their own credit-bearing projects or work directly with developers who would create dedicated projects for them. If marketable credits are produced by these projects, they should not be sold. They should be "retired."

Norfolk Southern's "Trees and Trains" program illustrates how a railroad can create its own carbon credits. Through Trees and Trains, Norfolk Southern partners with GreenTrees, a non-profit organization whose goal is reforesting 1 million acres in the Mississippi Alluvial Valley, the largest U.S. watershed.[350] While reforestation is a challenging mechanism for carbon offsetting,[351] this is an ambitious project with important multiple benefits. In addition to *carbon sequestration*—the removal of carbon dioxide from the atmosphere—forests release oxygen, prevent soil erosion, hold water to mitigate flooding, and provide wildlife habitat, recreational opportunities, and aesthetic enjoyment.

Descending Beaumont Hill, Palm Springs, CA. Photo by Craig Walker.

CHAPTER SEVEN

Railroad Environmental Sustainability

Let's conclude with a brief review of railroad sustainability reports, and a final thought about the role of railroads in creating a sustainable, clean energy future.

7.1 Railroad Sustainability Reports

The sustainability reports issued annually or biennially by the Class I freight railroads and Amtrak are impressive. They describe the serious commitments the railroads have made to their employees, the community and the environment—including commitments to save energy and reduce criteria pollutants and greenhouse gas emissions. To their credit, some reports also provide third-party-certified, transparent annual greenhouse gas inventories that are compared with company emissions reduction goals. Successes and failures in meeting those goals are openly discussed. These steps are consistent with a general trend of large and small companies in the United States. Forty-eight percent of Fortune 500 companies and 63% of Fortune 100 companies have set greenhouse gas emissions reduction goals and/or targets to increase renewable energy sourcing.[352]

Here are some energy and climate highlights from recent Class I railroad and Amtrak sustainability reports:

Amtrak[353]
- Implements its environmental and sustainability policies through a multi-layered organization of corporate officials and sustainability staff, and requires all new employees to take an environment and sustainability awareness course
- Exceeded diesel fuel conservation goal by 2.4 million gallons and saved 3 billion KWh of electricity with new ACS-64 electric locomotives over 2015-2016 two-year period

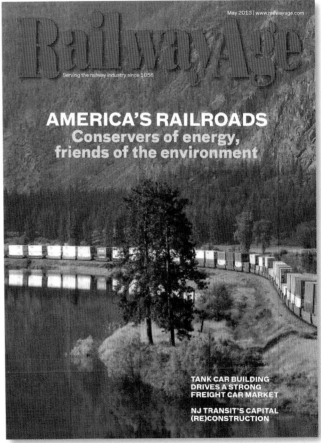

May 2013 issue of *Railway Age* highlighting railroad green efforts—an annual feature of the magazine. Image Credit: *Railway Age.*

BNSF[354]

- Achieved a 2.8% improvement in average corporate fuel efficiency between 2014 and 2015, and a 12% improvement between 2006 and 2015
- Equipped 95% of its locomotives with anti-idling technology by 2015 and investigated alternative locomotive fuel (LNG)

Canadian National[355]

- Achieved a 36% improvement in fuel economy (measured in gallons per gross ton-miles) between 1995 and 2015
- Reduced greenhouse gas "intensity" (GHG emissions/RTM) by 19% between 2005 and 2015

Canadian Pacific[356]

- Achieved 507 revenue ton-miles/gallon corporate fuel economy in 2013, which was 7% better than the Class I railroad average of 473
- While revenue ton-miles increased by 11% from 2010 to 2013, greenhouse gas emissions dropped by 3.4% during this period

CSX[357]

- Spent over $2.8 billion on fuel efficiency technology in last decade
- Established goals of reducing greenhouse gas emissions by 6% to 8% and increasing its fuel blend to 10% renewable fuel by 2020.

Kansas City Southern[358]

- Uses a cross-functional committee to set standards, publish guidelines, and monitor locomotive fuel consumption, train handling, and shutdown procedures to improve energy efficiency and conservation
- Increased average corporate fuel economy by 40% between 2005 and 2015

Norfolk Southern[359]

- Established goal of reducing annual fuel costs by $80 million between 2015 and 2020, which would require an 8.6% improvement in locomotive fuel economy
- Successfully saved 7.4 million gallons of diesel fuel and avoided 250,840 metric tons of GHG emissions in 2016

Union Pacific[360]

- Improved average corporate fuel economy by 32% between 2000 and 2010, with some recent fluctuations due to changing commodity mixes and business conditions
- Since 2000, invested $8 billion to purchase more than 4,300 new locomotives while retiring more than 3,000 locomotives that were less efficient and more polluting[361]

Some of the railroads have also quantified the GHG emissions reductions their customers have produced by shipping with them. BNSF, for example, reported that

its customers reduced total carbon dioxide emissions by 34 million tons in 2015 by shipping with BNSF instead of by truck. Based on EPA conversion factors, BNSF stated that this annual emission reduction was equal to taking 7 million vehicles off the road for a year or the amount of carbon sequestration that 27 million acres of trees accomplish in 1 year.[362] This is an important achievement, though a comprehensive assessment of the company's contribution to climate change should also consider the implications and consequences of BNSF coal-hauling. BNSF hauls more coal than any other railroad.[363] While the carbon footprint of coal-burning might be formally assigned to the power plants that burn coal, the mines that produce it and the railroads that haul it are part of the process, and, arguably, bear part of the responsibility.

Locomotive manufacturers like General Electric, Progress Rail's Electro-Motive Diesel, and Siemens demonstrate their commitments to energy efficiency and environmental responsibility daily by developing and manufacturing/remanufacturing advanced locomotives that are energy efficient and meet emissions regulations. These companies also have their own sustainability programs designed to conserve energy and reduce the environmental footprint of their own operations. For example:

- General Electric reports that it reduced its energy intensity and

WWII-era New York Central Railroad print advertisement boasting about the railroad's commitment to coal. Then, the NYC was hauling 95 million tons of coal annually. The ad refers to coal as *Black Magic* and *America's Fighting Fuel*, reflecting its importance at an earlier time when acid rain and climate change were not identified as issues. Image Credit: New York Central Railroad.

greenhouse gas emissions by more than 30% between 2005 and 2014.[364]

- Caterpillar, Progress Rail's parent company, reported a 17% reduction in energy use and 29% reduction in GHG emissions, respectively, between 2006 and 2016. The company also reported that 27% of its energy needs are now met by renewable energy.[365]
- Siemens has pledged to spend 100 million Euros to go carbon neutral by 2030.[366]

The Association of American Railroads' commitment to environmental excellence and sustainability is reflected in the organization's John H. Chafee Environmental Excellence Awards and its Professional Environmental Excellence Awards. These are given annually to railroad professionals who demonstrate industry-leading environmental achievement.

7.2 The Cargo Conundrum

Railroads are suited for hauling big, bulky, heavy things, and as such, seemed to have been made to haul coal. Steam locomotives devoured coal while transporting it to power the industrial revolution in the 19th and early 20th centuries. Coal has been called *Black Diamonds* and the *Lifeblood of American Railroading*.[367] The AAR reports that historically coal has been the most important commodity for U.S. railroads, and that in 2016 coal-hauling represented 31.6% of Class I freight rail tonnage and 13.9% of rail revenue.[368] However, coal is arguably the dirtiest fossil fuel, and a major contributor to climate change.

Coal remains a mainstay of electric power production (generating 30% of U.S. electricity in 2016[369]), but its use in that sector is in decline. Since 2008, over 200 U.S. coal-fired power plants have closed.[370] As the number of power plants burning coal has declined, so have rail shipments of coal—down 3.5 million carloads in 2016 compared to the peak coal shipment year

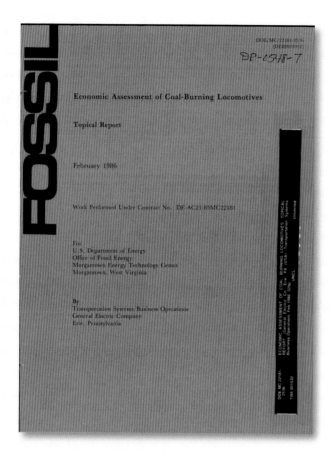

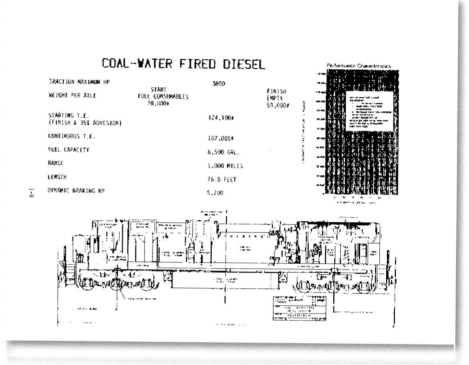

As late as 1986, a full 30 years after dieselization effectively ended the use of coal as locomotive fuel, GE proposed coal-burning diesel and gas turbine locomotives to the U.S. Department of Energy. Liquid coal slurry fuel would have fired these prime movers. The proposed locomotives were not built. Image Credits: U.S. Department of Energy.

2008.[371] AAR estimates that this reduction equals more than 30,000 fewer coal trains per year (based on an average of 115 carloads per coal train).

Both environmental regulation and lower natural gas prices (accompanied by an increase of natural gas-burning in power plants) have contributed to lower coal use. However, railroad analysts have said that natural gas has been more of a factor. If that is the case, then a switch to pro-coal government policies will not bring coal tonnage back to previous levels—though these policies (and, potentially, international shipments) can pause or slow the inevitable decline.[372] Structural changes in the U.S. and global economies—including increasing competition from wind and solar energy—are causing coal consumption to decline. This point has been increasingly echoed by railroad industry leaders. For example, the late Hunter Harrison explained in the July 19, 2017, *Financial Times*, "Fossil fuels are dead. That's a long-term view. It's not going to be in two or three years. But it's going away, in my view." He made this statement while running CSX, one of the nation's largest coal

Canadian National ES44DC locomotive leads a westbound stack train out of the Thompson River Canyon at Lytton, BC, in 2015. Soon the train will enter the famous Fraser River Canyon, the final major natural obstacle on the transcontinental mainline to the docks at Vancouver. Photo credit: Tony Schill.

Passenger and Freight Rail in a Low Carbon Economy

The recipe for a low carbon economy includes greater use of passenger rail, especially with highly efficient locomotives powered directly or indirectly by low/no-carbon energy. Passenger rail should thrive once society makes a serious commitment to address climate change.

Freight railroads can also thrive if they reposition themselves and find ways to survive the technological disruptions ahead. With today's technology, shipping freight by rail still makes the most energy sense. For that reason, it should be a national priority and goal of government policy to assist freight railroads, and not allow the destructive decline that might otherwise occur with the introduction of autonomous electric "super trucks" or unhelpful shifts in trade patterns. One positive development is that some railroads have already begun to specialize in hauling renewable energy systems. Union Pacific, for example, reports that in the last decade it has hauled the equivalent of 5,000 wind turbines. BNSF has also specialized in shipping wind turbines.[374]

haulers. Harrison previously was CEO of Canadian Pacific and Canadian National railroads. He was highly respected for his insights into the railroad industry.

What would American railroads look like with less reliance on coal—or, for that matter, other fossil fuel commodities that include oil and sand for hydrofracking oil and natural gas? Could they stay in business?

While there are no easy answers to these questions, at least one analyst has suggested that declining revenue from decreased coal shipments might be made up by additional cost cutting accompanied by more rail shipments of chemicals, intermodal, oil, manufacturing (e.g., auto and auto parts), and even fresh water as climate change takes hold.[373]

BNSF train hauling wind turbine blades, Laurel, MT. Credit: Billings Gazette.

In October 2018, the Intergovernmental Panel on Climate Change updated its scientific findings and projections regarding anthropogenic climate change. The report painted an even dire picture of current trends. It warned that unprecedented rapid changes to the world economy are needed to quickly reduce greenhouse gas emissions. Or we face worsening consequences of climate change and potential tipping points that could lead to runaway climate instability. The need to transition away from fossil fuels has never been greater. Railroads can play an increasingly constructive role.

A wholesale shift to a clean energy economy would provide significant economic opportunity for the freight railroads as well as the trucking industry. Imagine the volume of materials and products that would be needed to double the energy efficiency of every building in America, or to put solar panels and solar roofs on all those buildings. Or to install tens of thousands of additional wind turbines to shift the mix of generation toward zero carbon energy sources. An environmentally sustainable economy should be able to advance the most energy efficient means of moving commodities and people—the railroads.

Depew, NY, November 2014. Photo by author.

Acronyms and Abbreviations

AAR – Association of American Railroads
ABT – Average, Banking, and Trading
AESS – automatic engine start stop
APU – auxiliary power unit

BCM – Brake Cylinder Maintaining (brake valve)
BHP – brake horsepower
BHP-hr – brake horsepower-hours (a unit of energy or work)
BLET – Brotherhood of Locomotive Engineers and Trainmen
BNSF Railway – formerly Burlington Northern Santa Fe Railway
BSFC – brake specific fuel consumption
BTU – British thermal unit

CARB – California Air Resources Board
CCS – carbon, capture, and sequestration
CNG – compressed natural gas
CO – carbon monoxide
CO_2 – carbon dioxide
CTC – Centralized Traffic Control

DEF – diesel exhaust fluid
DOE – U.S. Department of Energy
DP – distributed power

ECP – electronically controlled pneumatic (brakes)
ECU – engine control unit
EGR – exhaust gas recirculation
EMD – General Motor's Electro-Motive Division or Progress Rail Services' Electro-Motive Diesel
EOT – end of train (device)
EPA – U.S. Environmental Protection Agency
ERAD – event recorder automatic download
FRA – Federal Railroad Administration

GE – General Electric
GF – gauge face
GHG – greenhouse gas
GM – General Motors
GTM – gross ton-mile
GTO – gate turn-off thyristor semiconductor
GWP – global warming potential

HC – (unburnt) hydrocarbons
HCCI – homogeneous-charge compression ignition
HEP – head-end power
HHV – higher heating value
HMI – human machine interface
hp – horsepower
hp-hr – horsepower-hour
HVAC – heating, ventilation, and air conditioning

IGBT – insulated gate bipolar transistor
IPCC – Intergovernmental Panel on Climate Change

KCS – Kansas City Southern Railway Company
kWh – kilowatt-hour

LEADER – New York Air Brake's Locomotive Engineer Assist/ Display & Event Recorder
LHV – lower heating value
LNG – liquid natural gas

mpg – miles per gallon
mph – miles per hour
MU – multiple unit
MWh – megawatt-hour

NBB – National Biodiesel Board
NEC – Amtrak's Northeast Corridor
NGO – non-governmental organization
NOx – nitrogen oxides
NTHP – net traction horsepower
NTSFC – net traction specific fuel consumption

PEMFC – polymer electrolyte membrane fuel cell
PM – particulate matter
PPA – power purchase agreement
ppm – parts per million
PRIIA – Passenger Rail Investment and Improvement Act of 2008
psi – pounds per square inch
PTC – Positive Train Control

RAILTEC – Rail Transportation and Engineering Center
R&D – research and development
REC – renewable energy certificate
rpm – revolutions per minute
RTM – revenue ton-mile

sCO_2 – supercritical carbon dioxide
SCR – selective catalytic reduction
SOFC – solid oxide fuel cell
SOFC-GT – solid oxide fuel cell with gas turbine
SOx – sulfur dioxide

T&D – transmission and distribution
TEG – thermoelectric generator
TGV – Train á Grande Vitesse (French high-speed train)
THP – traction horsepower
TOP – top of rail

UCS – Union of Concerned Scientists
UP – Union Pacific Railroad Company

WHRS – waste heat recovery system
WILD – wheel impact load detectors
WPT – wireless power transfer

ZEBL – zero emissions booster locomotive

Jeanette, PA, June 2015. Photo by author.

Selected Sources and Recommended Reading

Sources and recommended readings shown in underlined blue font below were available on the internet at the time this book was being finalized.

John H. Armstrong, *The Railroad: What It Is, What It Does* (Simmons-Boardman Books, Inc., Omaha, NE, 1998). See chapter on locomotives.

Association of American Railroads, "The Environmental Benefits of Moving Freight by Rail," June 2017.

Association of American Railroads, "Freight Railroads Help Reduce Greenhouse Gases," background paper, April 2016.

Association of American Railroads, "Railroads and Coal," background paper, July 2016.

Association of American Railroads, "US Rail Crude Oil Traffic," background paper, November 2015.

R. Barton and T. McWha, *Reducing Emissions in the Rail Sector: Technology and Infrastructure Scan and Analysis*, National Research Council Canada, September 2012.

Aviva Brecher, *Best Practices and Strategies for Improving Rail Energy Efficiency*, Office of Research and Development, Federal Railroad Administration, U.S. Department of Energy, January 2014.

California Air Resources Board, *CARB Draft Technology Assessment - Freight Locomotives*, April 2016.

Center for Transportation Research, *Railroad and Locomotive Technology Roadmap*, Argonne National Laboratory (Oak Ridge, TN: U.S. Department of Energy, 2002).

Classic Trains magazine, *Diesel Victory,* special edition (Kalmbach Publishing, Waukesha, WI, 2006).

Federal Railroad Administration, *Comparative Evaluation of Rail and Truck Fuel Efficiency on Competitive Corridors*, November 19, 2009.

Charles F. Foell and M.E. Thompson, *Diesel Electric Locomotive*, New York, NY (Diesel Publications, Inc), 1946.

W.N. Fritts, *Diesel-Electric Locomotives*, International Correspondence Schools-Scranton, PA, 1947 edition.

Sean Graham-White, *GE Evolution Locomotives* (Voyageur Press, 2007).

David L. Greene et al., "Reducing Greenhouse Gas Emissions from U.S. Transportation," Pew Center on Global Climate Change, January 2011.

William Hay, *Railroad Engineering* (John Wiley and Sons, 1953). See chapter on diesel-electric locomotives in more recent editions of this book.

Michael Iden, *Engines of Change & Future Fuels for US Freight Locomotives*, presentation to the Faster Freight Cleaner Air 2008 Conference, Los Angeles, CA, February 25-27, 2008.

Michael Iden, "Freight Railroad Energy: Alternatives and Challenges," presentation at RAILTEC Conference, University of Illinois at Urbana-Champaign, February 15, 2013.

Michael Iden, "2015 Motive Power Review," *Locomotive 2015*, special edition of *Trains* magazine (Kalmbach Publishing, Waukesha, WI, 2015).

Johannes Kech, "How Does Exhaust Gas Recirculation Work?," MTU Report, June 17, 2014. Note also available from Dr. Kech and MTU are these other excellent short white papers: "Turbocharging: Key Technology for High Performance," "How Does Selective Catalytic Reduction Work?," "How Does Common Rail Injection Work?," and "How Does a Diesel Particulate Filter Work?"

J. Parker Lamb, *Evolution of the American Diesel Locomotive* (Indiana University Press, Bloomington, IN, 2007).

Greg McDonnell, *Locomotives: The Modern Diesel & Electric Reference* (Boston Mills Press, Richmond Hill, Ontario, 2008).

William D. Middleton, *When Steam Railroads Electrified* (Kalmbach Publishing, Milwaukee, 1974 and 1976).

Bill Moyer et al., *Solutionary Rail: A People-Powered Campaign to Electrify America's Railroads and Open Corridors to a Clean Energy Future*, 2016. www.solutionaryrail.org; Solutionary Rail Video.

Karen Parker, *How a Steam Locomotive Works*, (TLC Publishing, Inc, Forest, Virginia, 2008). This excellent book is out of print though copies can occasionally be found on eBay or available through interlibrary loan, e.g. Cumberland County Public Library.

Chris Pinney and Brian Smith, "Cost Benefit Analysis of Alternative Fuels and Motive Design," Transportation Technology Center, sponsored by the Federal Railroad Administration, U.S. Department of Transportation, April 2013. *Progressive Railroading* magazine, www.progressiverailroading.com.

Rail Transportation and Engineering Center (RAILTEC), "Transitioning to a Zero or Near-Zero Emission Line-Haul Freight Rail System in California: Operational and Economic Considerations," University of Illinois at Urbana-Champaign, Spring 2016.

RailServe, www.railserve.com. A comprehensive guide to 19,000 railroad websites.

Railway Age magazine, www.railwayage.com.

Railway Educational Bureau, *Diesel Theory – Principles Explained* (Simmons-Boardman Books, Inc., Omaha, NE, 2007).

Railway Gazette, www.railwaygazette.com.

Mohammad Rasul et al., *Future Power Technologies - Final Report*, CRC for Rail Innovation, Australian Government Initiative, 2013.

Mark Reutter, *The Diesel Revolution,* special issue of *Railroad History*, Railway and Locomotive Historical Society, 2000.

Paul E. Rhine, *Fuel-Saving Techniques for Railroads: The Railroader's Guide to Fuel Conservation* (Simmons-Boardman Books, Inc., Omaha, NE, 2007).

Brian Solomon, *The American Diesel Locomotive* (MBI Publishing Company, Osceola, WI, 2000).

Brian Solomon, *Field Guide to Trains – Locomotives and Rolling Stock* (Voyageur, Minneapolis, MN, 2016).

Brian Solomon, *GE and EMD Locomotives* (Voyageur, Minneapolis, MN, 2014).

Brian Solomon, *Modern Locomotives: High Horsepower Diesels 1966-2000* (Crestline, New York, 2002).

Trains magazine (Kalmbach Publishing, Waukesha, WI).

Trains magazine, *Locomotive 2018* , special edition, (Kalmbach Publishing, Waukesha, WI). See other *Trains* locomotive annual issues as well.

Trains magazine, *Our GM Scrapbook,* (Kalmbach Publishing, Waukesha, WI, 1971)TRL and Ricardo, *Great Britain Rail Powertrain Improvements*, prepared for the UK Department of Transport, March 14, 2012.

William Vantuono, *All About Railroading* (Simmons-Boardman Books, Inc., Omaha, NE, 2006). See chapter "The Power that Moves Freight Trains."

Jeff Wilson, *Guide to North American Diesel Locomotives* (Kalmbach Publishing, Waukesha, WI, 2017).

Jeff Wilson, *The Model Railroader's Guide to Diesel Locomotives*, (Kalmbach Publishing, Waukesha, WI, 2009).

On Climate Change

Al Gore, *Truth to Power: An Inconvenient Sequel* (Rodale Press, New York, NY, 2017).

James Hansen et al., "Assessing 'Dangerous Climate Change': Required Reduction of Carbon Emissions to Protect Young People, Future Generations and Nature," PLOS ONE, December 3, 2013.

James Hansen, *Storms of My Grandchildren: The Truth about the Coming Climate Catastrophe and Our Last Chance to Save Humanity*, Bloombury USA, 2009.

Intergovernmental Panel on Climate Change, *Fifth Assessment Report*, 2013-2014.

Elizabeth Kolbert, *Field Notes from a Catastrophe: Man, Nature, and Climate Change*, 2015.

Michael Mann and Lee R. Kump, *Dire Predictions: Understanding Climate Change – Visual Findings of the IPCC* (DK Penguin Random House, New York, NY, 2016).

Joseph Romm, *Climate Change: What Everyone Needs to Know* (Oxford Press, New York, NY, 2016).

U.S. Global Change Research Program, *Climate Science Special Report – Fourth National Climate Assessment*, Wuebbles, D.J., D.W. Fahey, K.A. Hibbard, D.J. Dokken, B.C. Stewart, and T.K. Maycock, eds.), Washington, D.C., 2017.

World Wildlife Fund et al., "Power Forward 3.0: How the Largest Companies Are Capturing Business Value While Addressing Climate Change," 2017.

Videos of Interest

Big Power, *Trains* magazine (Kalmbach Publishing, Waukesha, WI, 2007).

Conrail/EMD SD80MAC Operator Familiarization, Railroad Video Productions, available from Anchor Videos, www.train-video.com.

Locomotive 2017: From Vintage to Modern, North America's Motive Power, *Trains* magazine (Kalmbach Publishing, Waukesha, WI, 2017).

Power & Speed: Diesel Locomotive, *Trains* magazine (Kalmbach Publishing, Waukesha, WI, 1997).

YouTube, www.YouTube.com (Numerous excellent diesel-electric locomotive videos are posted.)

Websites of Interest

American Railway Engineering and Maintenance-of-Way Association, www.arema.org., www.american-rails.com. See section on diesel locomotives.

Association of American Railroads, www.aar.org.

Brotherhood of Locomotive Engineers and Trainmen, www.ble-t.org.

Coordinated Mechanical Associations, www.rsiweb.org/cma.

Diesel Era magazine, www.dieselera.com.

Diesel Locomotive Technology, www.railway-technical.com/trains/rolling-stock-index-l/diesel-locomotives

Dieselnet, www.dieselnet.com.

The Diesel Shop, www.thedieselshop.us.

Federal Railroad Administration, www.fra-dot.gov.

Index to Railroad Historical Societies, http://www.rrhistorical.com/rrpro/database.html.

Locomotive Wiki, http://locomotive.wikia.com/wiki/Locomotive_Wiki.

Midwest High Speed Rail Association, www.midwesthsr.org., *Model Railroader* magazine, www.mrr.trains.com

Rail Passengers Association, www.railpassengers.org.

National Railway Historical Society, www.nrhs.org.

Norfolk & Western Historical Society, www.nwhs.org.

Passenger Train Journal magazine, www.passengertrainjournal.com

Pennsylvania Railroad Technical & Historical Society, www.prrths.com.

Progressive Railroading, www.progressiverailroading.com.

Railfan and Railroad magazine, www.railfan.com

Rail Passengers Association, www.railpassengers.org.

Railroad Model Craftsman magazine, www.rrmodelcraftsman.com

Railroad Workers Union, www.railroadworkersunited.org.

Railvolution, www.railvolution.net/railvolution.

Railway Age, www.railwayage.com., www.transalert.com.

Railway Gazette International, www.railwaygazette.com.

Railway Supply Institute, www.rsiweb.org.

Railway Technical Website, www.railway-technical.com.

Railway Technology Website, www.railway-technology.com.

Steel Interstate Coalition, www.steelinterstate.org.
Steve's Railroad Pages, http://members.localnet.com/~docsteve/railroad/en_info.htm.

Trainorders.com, www.trainorders.com.

Trains magazine, http://trn.trains.com

Website for Energy Efficiency Technologies for Railways, http://www.railway-energy.org.

Wikipedia is an excellent source of information about diesel-electric locomotives, component technologies, and energy issues.

Amtrak and Class I Railroad Websites

Amtrak, www.amtrak.com

BNSF, www.bnsf.com

Canadian National Railway, www.cn.ca

Canadian Pacific Railway, www.cpr.ca

CSX, www.csx.com

Kansas City Southern, www.kcsouthern.com/en-us

Norfolk Southern, www.nscorp.com

Union Pacific, www.up.com

Locomotive and Trainset Manufacturers

Alstom, www.alstom.com

Bombardier, https://www.bombardier.com/en/transportation.html

Brookville, www.brookvillecorp.com

GE Transportation, www.getransportation.com

MotivePower, www.wabtec.com

NRE, www.nre.com

Progress Rail Services Corporation - Electro-Motive Diesel, www.progressrail.com

Railpower, www.rjcorman.com

Siemens, www.siemens.com

Talgo, www.talgo.com

Rural Retreat, VA, July 2011. Photo Credit: Hugh Hopkins.

About the Author

Walter Simpson is an energy professional and environmental educator who served for 26 years as Energy Officer at the State University of New York at Buffalo (UB). At UB, he led a nationally recognized campus energy conservation program credited with over $100 million in savings. He also founded and directed the UB Green Office, UB's environmental sustainability office, and taught many college-level energy and environmental courses.

After studying engineering at Lehigh University, Walter received a B.A. in philosophy from Boston University, and master's degrees in philosophy and environmental studies from UB. He is editor/co-author of *The Green Campus: Meeting the Challenge of Environmental Sustainability* and author of *Cool Campus! A How-to Guide for College and University Climate Action Planning.* A former director of the Western New York Peace Center, he first became interested in the energy issue when he realized that energy waste and foreign energy dependence could cause war as well as environmental harm.

Learning about locomotive operation at the control stand of a GE Dash 9. Photo Credit: Paul Schiff

Walter has had a lifelong interest in railroads, and, with the assistance of many experts, has applied his knowledge of energy to the subject of diesel-electric locomotives—producing this groundbreaking, comprehensive volume. He and his wife Nan reside in a super-insulated solar home in Amherst, New York. Their son Jay works in law enforcement and their daughter Skye is a writer and editor.

The author can be reached through his www.diesel-electric-locomotives.com website which has been created to complement this book. Among other things, the website provides "live links" for the numerous articles, reports, and other resources referenced in this book's appendix and endnotes.

Toronto, Ontario, July 2016. Photo by author.

Notes

These notes are deliberately formatted title first (instead of author first) to make them more user-friendly to interested readers who want to dig deeper into the book's various topics. A lot of research went into this book. The endnotes are here to help you make use of it.

Many endnotes reference older issues of various railroad magazines such as *Trains* and *Railway Age*. Interested readers can often obtain copies of these magazines on eBay or at train shows.

The references shown in the endnotes in underlined blue font were available on the internet at the time this book was being finalized. They should be accessible through Google or another search engine if they are still posted on the web at the time of the search. Reference librarians, document authors, or the author of this book may be able to help with documents no longer available through web search. Walter Simpson can be reached via www.diesel-electric-locomotives.com.

---

1. U.S. Class I freight railroads are determined by annual operating revenues, now in excess of approximately $450 million. The value floats because of inflation. The seven Class I freight railroads are: BNSF Railway, CSX Transportation, Grand Trunk Corporation (Canadian National), Kansas City Southern Railway, Norfolk Southern Combined Railroad Subsidiaries, Soo Line Railroad (Canadian Pacific), and Union Pacific Railroad. Amtrak operating revenues place it within the Class I railroad definition.

2. *Railroad Facts, 2016 edition*, Association of American Railroads, 2016, page 63.

3. See www.thedieselshop.us data sheets for the ES44AC and SD70ACe.

4. "How the Diesel Changed Railroading," Jerry A. Pinkepank, *Diesel Victory, Classic Trains* magazine special edition (Kalmbach Publishing), 2006, pages 8-19; and "Why Dieselize," Brian Solomon, *The American Diesel Locomotive*, MBI Publishing Company, 2000, pages 9-17.

5. *The Steam Locomotive: Its Theory, Operation, and Economics Including Comparisons with Diesel-Electric Locomotives*, Ralph P. Johnson, New York, NY, and Omaha, NE (Simmons-Boardman Books, Inc.), 1942 (1981 edition), pages 437-439.

6. The Pennsylvania Railroad's 4-4-6-4 Q2 locomotive achieved 7,987 horsepower. The Chesapeake and Ohio's 2-6-6-6 H-8 "Allegheny" measured 7,500 hp.

7. Note that there are differences between tractive effort or force, horsepower, and efficiency *at the rails* vs. *at the drawbar*. Drawbar measurements are always lower than those measured at the rails because the former factor in the additional energy losses incurred by moving the weight of the locomotive.

8. "Where the Energy Goes," www.fueleconomy.gov, U.S. Department of Energy website.

9. *Diesel-Electric Locomotive*, Charles F. Foell and M.E. Thompson, New York, NY (Diesel Publications), 1946, page 19.

10. Some steam locomotives could achieve availability close to their diesel challengers. Paul Kiefer, New York Central's Chief Engineer for Motive Power, stated that NYC's 4-8-4 Niagara steam locomotive demonstrated 83% availability vs. 86% availability by an equivalent diesel-electric locomotive during a 17-day test period in October of 1944. Kiefer predicted annual 66% availability for the Niagaras vs. 73% for the diesel-electrics. See *A Practical Evaluation of Railroad Motive Power*," P.W. Kiefer, Steam Locomotive Research Institute – New York, 1947, pages 44-45.

11. *Diesel-Electric Locomotive*, Charles F. Foell and M.E. Thompson, page 15.

12. The U.S. Department of Transportation's Bureau of Transportation Statistics reports that U.S. railroads operated 24,707 locomotives that hauled freight 500 million miles between terminals (i.e., excluding yard work) in 2012. Thus, the average locomotive

traveled 20,237 miles/year. Given a 20- to 25-year lifespan (which can be extended by rebuilding), the average diesel-electric locomotive hauled freight 20,237 miles/year x 20 years = 404,744 miles over the course of its lifespan. A historical example of a diesel-electric locomotive whose lifetime mileage far exceeded this mean was Union Pacific's EMD DD40X which averaged 22,000 miles per month and within just 5 years had operated 1 million miles. A current example of a million-mile-plus locomotive is the Amtrak GE Genesis P42 locomotives. These have logged in excess of 2 million miles, as per "The Locomotive Everyone Loves to Hate (updated April 20)," Fred Frailey, *Trains* magazine blog, April 8, 2011.

13. "N&W Class J 611: The Spirit of Roanoke," Fireup611.org.

14. The compression ratio of a piston engine is defined as the ratio between the maximum and minimum volumes displaced by the engine's pistons during a full stroke.

15. For a brief authoritative discussion of the history of diesel-electric locomotives see the *Guide to North American Diesel Locomotives*, Jeff Wilson (Kalmbach Publishing, Waukesha, WI), 2017, pages 6-26, which includes an excellent visual history of diesel-electric locomotive development through 1960 by Greg McDonnell entitled "Timeline to Victory: How Diesels Won the Battle vs. Steam from 1905 to 1960." Also see *Diesel Victory* (Kalmbach Publishing), 2006.

16. See "Doodlebugs, the Popular Motorized Railcar," American-Rails.com.

17. "Herman Lemp – He Harnessed the Diesel," Carl Byron, *Diesel Victory* (Kalmbach Publishing), pages 46-48.

18. "The Diesel That Did It," *Confessions of a Train Watcher – Four Decades of Railroad Writing by David P. Morgan*, George H. Drury, Editor, Waukesha, WI (Kalmbach Publishing), 1997, page 34. (This article previously published in *Trains* magazine, February 1960.)

19. "1939's Big Breakthrough: the FT," *Trains* magazine, November 2014. Also see "The Most Famous Face in Railroading," J. David Ingles, *Trains* magazine, August 1996, page 48; and "F Means Freight," David P. Morgan, *Our GM Scrapbook*, *Trains* Magazine special issue, 1971, page 48.

20. "North American Locomotive Review 2016," Railinc. Also see "Advanced Powertrain Technologies and the North American Locomotive Market," Frost & Sullivan, June 6, 2011.

21. "GM vs. GE," Steve Glischinski, *Trains* magazine, September 1997, page 82. Also, "The Evolution of the American Diesel Locomotive," J. Parker Lamb, 2007, Bloomington, IN (Indiana University Press), page 172.

22. "GE to Sell or Spin Off Its Transportation Business," www.progressiverailroading.com, November 11, 2017; "Nothing Runs Like a . . . GE?," Steve Sweeney, *Trains* magazine, February 2018; and "An Upside Down World," Bill Stephens, *Trains* magazine, February 2018.

23. "Wabtec to Merge with GE Transportation," *Railway Gazette*, May 21, 2018.

24. "Charger Diesel-Electric Locomotive," Siemens.com/mobility.

25. "Electro-Motive F125 Fast, Safe, and Clean," EMD sales brochure; and "Siemens Charger: Clean Diesel-Electric Locomotives for Better Reliability and Efficiency," and "Charger Diesel-Electric Locomotive – All Aboard Florida" Siemens sales brochure.

26. For excellent brief, historical discussion of this topic, see chapter entitled "How Diesel Locomotives Work" in *Guide to North American Diesel Locomotives*, pages 27-36.

27. For an interesting series of presentations on the development of diesel-electric locomotive cabs and current challenges, see "The Future Locomotive: How to Manage What You Have Today With a View to the Future," proceedings from a July 30-31, 2013 conference sponsored by the Transportation Research Board of the National Academies of Science.

28. See "Positive Train Control" on the Association of American Railroads and the Federal Railroad Administration websites.

29. "Cab Technology Integration Lab Heads-Up Display Survey," FRA-2017-0002-0011, Federal Railroad Administration website.

30. "Next Generation Locomotive Cab – Task 5 Summary Report: Control Stand Design," Stephen J. Reinach and Abdullatif K. Zaouk, Foster-Miller, Inc., under the auspices of the Federal Railroad Administration, U.S. Department of Transportation, July 2010.

31. "Train Performance Improves with GE Transportation's Digital Solutions Software," GE Transportation.

32. "Canadian Pacific Selects GE's Trip Optimizer for Significant Locomotive Fuel Savings," Business Wire, July 9, 2009.

33. "Rail Connect 360," GE Imagination at Work, GE Transportation website. Also "Explore GE Transportation's Digital Solutions for the Rail Ecosystem," GE Transportation website.

34. "Locotrol Distributed Power System," GE Transportation.

35. "LEADER – The Train Handling and Energy Management System," LEADER product brochure, New York Air Brake website.

36. "Follow the LEADER," Norfolk Southern 2014 Sustainability Report. Also see "Follow the LEADER" in Norfolk Southern's 2016 Sustainability Report, which states that by the end of 2015, nearly 75% of NS's road fleet was equipped with LEADER.

37. "LEADER – The Train Handling and Energy Management System.

38. "We Help Our Customers Save Money," Fuel Saving Tools, Wi-Tronix website.

39. "Greening the Railroads and Why It Matters," William Vantuono, Railway Age, January 31, 2011.

40. "Locomotive Event Recorders," 49 CFR 229, Federal Railroad Administration, U.S. Department of Transportation, page 37941 (Docket No. FRA-2003-16357, Notice No. 3).

41. "PTC System Information," Federal Railroad Administration, U.S. Department of Transportation.

42. "Positive Train Control," Association of American Railroads, June 2016.

43. Ibid.

44. As a general rule, EMD fuel tanks are bolted to the frame, while GE fuel tanks are welded to it – and GE passenger locomotives have integral frames. Also see "Four Things to Know About Fuel," Chris Guss, Trains magazine, November 2017, page 18. Dated but still very informative about diesel fuel, "What You Should Know About Diesel Diets," Dibrell Du Val, Trains magazine, May 1973, page 36.

45. "Four Things to Know About Fuel," Chris Guss, Trains magazine, November 2017, page 18.

46. The SpillX Locomotive Fueling System, www.spillx.com; and Direct-to-Locomotive Fueling, Locomotive Service, Inc.

47. Dated but still of interest is "What You Should Know About Diesel Diets," Dibrell Du Val, Trains magazine, May 1973, pages 36-40.

48. "Heat of Combustion," www.wikipedia.org.

49. "Fuel Specifications and Pricing," Union Pacific website.

50. Railroad Facts, 2016 edition, page 63.

51. "Union Pacific Reports Fourth Quarter and Full Year 2016 Results," Union Pacific website, January 19, 2017.

52. "Union Pacific – 2016 Building American Report," Union Pacific website, page 35.

53. *GE Evolution Locomotives*, Sean Graham-White (Voyageur Press, St Paul, MN), page 46.

54. "Turbocharging: Key Technology for High-Performance Engines," Johannes Kech, Ronald Hegner, and Tobias Mannle, MTU White Paper, 2014.

55. See "APU (Auxiliary Power Unit)," HOTSTART website, for tables detailing exact fuel consumption amounts for various EMD and GE locomotives when idling.

56. See 40 CFR Part 1033.530 for the duty cycle used by U.S. EPA in emissions calculations. Presumably, this 38% idling time is based on field measurements and is pre-adoption of anti-idling locomotive technologies.

57. "Locomotive Idling Reduction," David E. Brann (presentation at National Idling Reduction Planning Conference, May 2004).

58. "SmartStart IIe," brochure, ZTR. www.Ztr.com.

59. "HOTSTART APU Locomotive Idle Reduction System," YouTube video.

60. "Just the Basics – Diesel Engines," Freedom Car and Vehicle Technology Program, U.S. Department of Energy, August 2003.

61. *GB Rail Powertrain Improvements*, TRL and Ricardo consultants, prepared for the UK Department of Transport, March 14, 2012.

62. For a thorough discussion of HCCI engine technology and its potential application to locomotives, see U.S. Department of Energy *Railroad and Locomotive Technology Roadmap*, pages 34-38.

63. For automotive application, see "Thermoelectric Conversion of Exhaust Gas Waste Heat into Usable Electricity," Gregory P. Meisner, General Motors Global Research & Development, paper presented at 2011 Directions in Engine-Efficiency and Emissions Research (DEER) Conference, Detroit, Michigan, October 5, 2011.

64. *Turbines Westward*, Thos. R. Lee (T. Lee Publications, Clay City, KS), 1975.

65. "How High the Horsepower," Paul D. Schneider, *Trains* magazine, September 1996, pages 34-41 and "Crossing the 6,000-hp Threshold," Gus Welty, *Railway Age*, July 1996, pages 33-36. Also "The 6,000 Horsepower Milestone," by Sean Graham-White in *The American Diesel Locomotive*, Brian Solomon, pages 148-149.

66. SD70MAC Systems/Operations Manual, GM Electro-Motive Division, 1993, pages 1-5 (as provided to Burlington Northern Railroad).

67. Prior to the advent of solid-state inverters, it was difficult to control the speed of AC motors while still maintaining the torque needed for traction purposes.

68. A "flashover" is the discharge or arcing of high-voltage positive polarity electricity to ground. Carbon dust from worn commutator brushes in DC motors can be the cause.

69. "N&W's Secret Weapons," Robert A. Le Massena, *Trains* magazine, November 1991, pages 64-69. The Y6c was a variant of the Y6b that Le Massena says received a "booster valve," which introduced higher temperature superheated steam into the locomotive's larger compound cylinders, and other improvements.

70. Hysteresis and eddy current losses apply to power losses in an inductor, where the former occurs when a core is magnetically deformed by expanding and contracting magnetic fields and the latter occurs when stray currents are induced and lost in the core.

71. See "EMD, GE Get Competitive: Six Axle Units with Only Four AC Traction Motors Is Big Test," *Trains* magazine, May 2013, page 16.

72. "Freight Locomotive Trucks," Chris Guss, *Trains* magazine, July 2017, page 18.

73. "GE Transportation Unveils New Evolution Series Locomotive," William C. Vantuono, *Railway Age*, May 18, 2009.

74. _Comparative Evaluation of Rail and Truck Fuel Efficiency on Competitive Corridors_, U.S. Federal Railroad Administration, November 19, 2009, page 28.

75. Ibid.

76. See EMD presentation "The Merits of AC vs. DC Locomotives," November 13, 2008, slide 14. "Dispatchable adhesion" is the amount of adhesion dispatchers can count on with 99% confidence from a given locomotive in all weather circumstances.

77. The development of hi-adhesion trucks began as early as the 1960s, with EMD partnering with an Australian locomotive builder to offer a first generation in 1964. EMD tested its prototype HT-C and GH-C trucks in 1970 on seven SD-45X experimental locomotives. See "Design Features and Performance Characteristics of the High Traction, Three-Axle Truck," H.A. Marta, K.D. Mels, and G.S. Itami, ASME-IEEE Joint Railroad Conference, March 14-15, 1972.

78. _Diesel Electric Locomotive_, Charles F. Foell and M.E. Thompson, pages 110 and 135.

79. _SD40-2 Operator's Manual_, Electro-Motive Division of General Motors, LaGrange, IL, March 1980, pages 3-21.

80. "Merits of AC vs DC Locomotives," GM Electro-Motive Division, November 13, 2008, slide 22.

81. "This Software-Guided Supersonic Air Blower Sweeps the Rails Clean," Greg Petsche, GE Reports, April 27, 2015.

82. "Watching Your Weight," Chris Guss, _Trains_ magazine, April 2018, page 18.

83. See, for example, "Friction Management" tab on L.B. Foster Rail Technologies website.

84. "Goop Be Gone," Mischa Wanek-Libman, _Railway Age_, February 2017, pages 34-38.

85. "Our Key Energy Initiatives," Norfolk Southern _2013 Sustainability Report_, page 24, and "Evaluation of a Top-of-Rail Lubrication System," Research Results, Federal Railroad Administration, U.S. Department of Transportation.

86. "Braking Systems -- Pumped Up about High Tech," Stuart Chirls, _Railway Age_, December 2017; "NYAB's DB-60 II Control Valve with Brake Cylinder Maintaining," New York Air Brake video; and "Increasing Safety & Productivity Now by Minimizing the Impacts of Brake Cylinder Leaks," New York Air Brake webinar hosted by _Railway Age_, November 10, 2015.

87. "End of Train Devices," Robert S. McGonigal, _Trains_ magazine website, May 1, 2006.

88. "State-of-the-Art EOT," _Railway Age_, October 2017, page 36.

89. "Evaluation of Rail and Truck Fuel Efficiency on Competitive Corridors," Federal Railroad Administration, November 19, 2009, page 25.

90. "Breakthrough in Braking," Greg McDonnell, _Trains_ magazine, November 1996, pages 70-77.

91. "DOT Fails to Make Convincing Case for ECP Braking Technology, GAO Report Shows," Association of American Railroads, October 12, 2016. Also see "US DOT Repeals ECP Brake Rule," William Vantuono, _Railway Age_, December 5, 2017.

92. "Feasibility of Load-Shedding to Improve Efficiency and Reduce Energy Consumption," Melissa Shurland et al., March 23-26, 2015, Joint Rail Conference, San Jose, CA, Rail Transportation Division, American Society of Mechanical Engineers.

93. "Demand Control Ventilation," Building Technologies Program, U.S. Department of Energy.

94. "AMD103: Powering Amtrak into the 21st Century," David C. Warner, _Trains_ magazine, June 1993, pages 16-25.

95.   *Railroad and Locomotive Technology Roadmap*, page 34.

96.   "Fuel-Electric," Land Speed Record for Rail Vehicles, Wikipedia, retrieved December 30, 2017.

97.   "China's Bullet Trains Just Became the World's Fastest," *Trains* magazine weekly newsletter, August 25, 2017.

98.   "French National Railroads Top World's Rail Speed Record for a Second Time – 207 mph," *Locomotive Engineers Journal*, May 1955. This record was set on March 29, 1955, by the BB 9004 electric locomotive which achieved 12,000 hp at maximum top speed. The article reports that the locomotive required 50,000 hours of research and 40,000 hours to build.

99.   "Engines of Change: Low Emissions 'Gen-Set' Switching Locomotives for Los Angeles Rail Yards," Mike Iden, presentation at the Faster Freight Cleaner Air Conference, Long Beach, CA, January 30 – February 1, 2006. See slide 7.

100. See "Gensets Get to Work," Greg McDonnell, *Locomotive 2007*, *Trains* magazine special edition, pages 22-27.

101. AAR's Chronology of Railroading in America, page 5. Also "Genset," Technology page of Environment Management section of Union Pacific Railroad website.

102. RJ Corman RP Product Line, NRE N-Vironmental 2GS-14B ULEL Switcher, and Railserve Leaf.

103. "Genset," Union Pacific website.

104. "Norfolk Southern Locomotive Fleet Overview," May 7, 2014. See "Genset Experiment," slides 16-17.

105. "The Engines That Have No Engines," David Lustig, *Trains* magazine, November 2010.

106. "Mother Units Could Love," David Lustig, *Trains* magazine, January 2012.

107. "The Locomotive Landscape," William C. Vantuono, *Railway Age*, September 9, 2015.

108. "2015 Locomotive Rebuilds," *Locomotive 2016*, *Trains* magazine special edition.

109. "Good as New: Locomotive Manufacturers Focus on Modernization Work as Demand for Units Dwindles," Daniel Niepow, *Progressive Railroading*, January 2018, pages 29-34.

110. "Norfolk Southern's SD60Es," *Diesel Era*, September/October 2014, pages 31-34. Also see "Green Is the New Black," Chris Guss, *Trains* magazine, March 2016, pages 30-37.

111. "Progress Rail Services and Electro-Motive Diesel Locomotive Emissions Webinar," Mike Klabunde and Jeff Moser, EPA Region 5 Webinar, October 27, 2010 (EPA Archived Document).

112. Ibid.

113. "DC2AC; Transforming DC Dash 9s into High Performance AC Machines," Greg McDonnell, *Locomotive 2016*, *Trains* magazine special edition.

114. "Electro-Motive 710ECO Repower Locomotive," Progress Rail website.

115. "Size Matters," Steve Bradley, *Locomotive 2016*, *Trains* magazine special supplement, pages 26 and 27.

116. "Progress Rail Services and Electro-Motive Diesel Locomotive Emissions Webinar" presentation, Mike Klabunde and Jeff Moser, October 27, 2010 (EPA Archived Document).

117. "Progress Rail, PHL to test Tier 4 Switch Locomotive," Progressive Rail press release, February 24, 2017; "Progress Rail Completes Emissions Testing for EMD24B, Tier 4 Switch Locomotive,"

Progress Rail press release, February 22, 2017, and "EMD24B Repower-T4," Progress Rail website.

118. "Locomotive Gems from GEMS," William Vantuono, *Railway Age*, March 20, 2018.

119. 2007 Association of American Railroads John Chaffe Award Nomination Form for Donald L. Robey.

120. Information from conversation with Norfolk Southern locomotive manager, September 14, 2017.

121. "The Environmental Benefits of Moving Freight by Rail," Association of American Railroads, June 2017.

122. "Comparative Evaluation of Rail and Truck Fuel Efficiency on Competitive Corridors," page 4.

123. "Sensitivity of Freight and Passenger Rail Fuel Efficiency to Infrastructure, Equipment, and Operating Factors," Garret A. Fullerton, Giovanni C. DiDomenico, and Tyler Dick, Transportation Research Board, 2015.

124. "AAR Standard Reference Conditions" from AAR Recommended Practice RP-589 (2001): elevation 1000 feet, air temperature 60 degrees F, fuel temperature 60 degrees F, specific gravity of fuel 0.645 (weight = 7.043 lb/gal), and fuel higher heating value 19,350 BTU/lb. RP-589 is expected to be updated soon.

125. "Transitioning to a Zero or Near-Zero Emission Line-Haul Freight Rail System in California: Operational and Economic Considerations," Rail Transportation and Engineering Center (RAILTEC), page 51.

126. General Electric Transportation states that its "products use up to 5% less fuel and are 6% more fuel efficient than GE Transportation's closest competitor in North America as validated by a nationally recognized, independent research institute." See "Increased Fuel Efficiency," Building Industry Leading Locomotives, GE Transportation website.

127. The 48% efficiency number is based on numerous conversations with locomotive experts and detective work by the author – in the context of non-disclosure agreements that prevented the sharing of hard data on this subject.

128. 200 hp is given as auxiliary load, consistent with published accounts, though an ES44AC schematic shows auxiliary nameplate horsepower totaling 344 horsepower (radiator fan motor – 122 hp; two air-to-air fan motors – 2 x 16 hp = 32 hp; alternator blower – 47 hp; air compressor motor – 63 hp; and traction blower motor – 80 hp). Equipment may be oversized and may not operate at full power when maximum efficiency is achieved.

129. "Transitioning to a Zero or Near-Zero Emission Line-Haul Freight Rail System in California: Operational and Economic Considerations," Rail Transportation and Engineering Center (RAILTEC), page 51.

130. Ibid., page 50.

131. "Specification for Diesel-Electric Locomotives," PRIIA Specification 305-005/Amtrak Specification No. 982, Revision A, July 10, 2012, pages 11, 117, and 271. Also see PRIIA Section 305 Committee website.

132. *Railroad and Locomotive Technology Roadmap*

133. U.S. DOE 21st Century Locomotive Technology program, Technical Status Report, 2003.

134. "21st Century Locomotive Technology (Locomotive System Tasks)," Robert D. King and Lembit Salasoo, GE Global Research, April 2006.

135. See, for example, "Roadmap to Energy Conversion Efficiency at Cummins," Wayne Eckerle, Cummins VP Corporate Research and Technology (presentation at ECR Symposium, Energy Research Center, University of WI at Madison, June 2013); and "Overview of High-Efficiency Engine Technologies," Wayne Eckerle (presentation at Directions in Engine Efficiency and Emissions Research Conference, October 2011).

136. "Disruption: The Future of Freight," Oliver Wyman, 2017.

137. *Railroad and Locomotive Technology Roadmap*, page 27.

138. *Fuel-Saving Techniques for Railroads – The Railroader's Guide to Fuel Conservation*, Paul E. Rhine (Simmons-Boardman Books, Inc., Omaha, NE), 2007, pages 58-59.

139. All locomotives in the consist must be equipped with the necessary technology, which must also be properly maintained, in order to implement this fuel-saving measure.

140. *Fuel-Saving Techniques for Railroads – The Railroader's Guide to Fuel Conservation*, Paul E. Rhine, page 85.

141. "Running Slow to Get Ahead," David Ibata, *Trains* magazine, October 2016, pages 6-7.

142. Ibid.

143. Union Pacific and BNSF, respectively; the UP program has been discontinued.

144. *Comparative Evaluation of Rail and Truck Fuel Efficiency on Competitive Corridors*, page 30.

145. Conversation with Norfolk Southern executive, December 2017.

146. Fuel Efficient Operating Practices and Processes," *BNSF Railway* magazine, Spring 2009, page 5.

147. "Union Pacific Saves Fuel While Increasing Efficiency," Union Pacific website.

148. *2014 BNSF Sustainability Report*, page 8.

149. "Fuel Efficient Operating Practices and Processes," BNSF Railway magazine, Spring 2009, page 5.

150. An interesting description of the Trip Optimizer is available in a 2009 progress report GE submitted to the U.S. DOE, "21st Century Locomotive Technology: Quarterly Technical Status Report 28 DOE/AL68284-TSR28," U.S. DOE Office of Scientific and Technical Information (OSTI), www.osti.gov, page 5.

151. "Beat the machine" means operate their locomotive or consist of locomotives with fuel economy better than what would have been achieved if Trip Optimizer or LEADER had been engaged.

152. "BLET Petitions FRA for Immediate Halt to LEADER, Trip Optimizer Technology," Brotherhood of Locomotive Engineers and Trainmen, February 5, 2016. Also see detailed February 4, 2016, letter to FRA's Robert Lauby.

153. This same problem is being encountered with automobiles equipped with automation features. Early studies are showing that they make people worse drivers. "As Robots Take Wheel, Driving Skills Begin to Hit the Skids," Keith Naughton, *Bloomberg News*, August 10, 2017.

154. This June 17, 2016, Federal Railroad Administration letter to the Brotherhood of Locomotive Engineers and Trainmen (as well as the International Association of Sheet Metal, Air, Rail and Transportation Workers Transportation Division) is in the author's possession but not posted on the FRA website.

155. "Will Computers Learn to Fly Well Enough Before Pilots Forget How?," Robert Graves, *This Is Your Captain Speaking*, www.robertgraves.com.

156. "World Report on Metro Automation," Union Internationale des Transports Publics (International Association of Public Transport), July 2016.

157. "Rio Tinto Going Driverless Down Under," *Railway Age*, August 2017 and "Rio Tinto's Automated Trains Approved," *Railway Gazette*, May 18, 2018.

158. "Whither the Train Crew?," Chris Jackson, *Railway Gazette*, August 7, 2016.

159. "Automatic Train Operation to Be Tested on Dutch Freight Corridor," *Railway Gazette*, January 22, 2018; and "Alstom to Participate in Dutch ATO Trains," David Briginshaw, *International Railway Journal*, January 22, 2018.

160. "Self-Driving Freight in the Fast Lane," Jason Kuehn and Juergen Reiner, in the Oliver Wyman Risk Journal, Vol. 5, 2017; "Disruption: the Future of Freight Rail," Oliver Wyman; and "Trains without Crews: Fantasy or the Future," Justin Franz, *Trains* magazine, January 2018, pages 26-33.

161. "Automation in the Railroad Industry," Federal Railroad Administration, U.S. Department of Transportation, Request for Information, March 29, 2018.

162. "Railroaders, Public Speak Out Against Automated Trains," Justin Franz, *Trains* magazine, April 24, 2018.

163. "Elon Musk Reveals the Dream Truck: Tesla Semi," www.youtube.com; "Tesla Unveils an Electric Rival to Semi-Trucks," Neal E. Boudette, *New York Times*, November 16, 2017; "Tesla Ups the Ante Once Again," David Nahass, *Railway Age*, December 2017, page 19; and "Tesla's Electric Semi Promise – But Will It Deliver?," Cody Cottier, D-Brief, *Discover Magazine*, November 17, 2017.

164. "Autonomous Trucks: An Elephant in the Economy," J. William Vigrass, *Railway Age*, March 2018; and "Tractor Trailers Without a Human at the Wheel Will Soon Barrel onto Highways Near You. What Does This Mean for the Nation's 1.7 Million Truck Drivers?," David H. Freedman, *Technology Review*, March/April 2017.

165. "Railtrends 2017 Recap," Tony Hatch, *Progressive Railroading*, December, 2017; and "Self-Driving Trucks Will Require Railroads to Take Next-Level Approaches to Digitization, Customer Service," Michael Popke, *Progressive Railroading*, January 2018.

166. "CSX & the Environment," CSX Transportation, October 21, 2008.

167. "Statement of Stephen Bruno, Vice President, Brotherhood of Locomotive Engineers and Trainmen, before the National Transportation Safety Board, Forum on Positive Train Control," February 27, 2013.

168. See LocoVISION product described in "Train Performance Improves with GE Transportation's Digital Solutions Software," GE Transportation; "In-Cab Cameras," *Union Pacific 2015 Building America Report*, page 17; "CP Urges Expansion of Locomotive Video/Voice Recorders," *Railway Age*, September 2016; and RailView: Locomotive Digital Video Recorder, Leidos.com.

169. "Aerodynamic Pseudocontainer for Reducing Drag Associated with Stacked Intermodal Shipping Containers – US Patent # 8,511,236 B2, Michael Iden, August 20, 2013.

170. "Union Pacific Unveils New Aerodynamic Technology for Double-Stack Intermodal Trains," Union Pacific website.

171. "2-Mile Trains Trending," Chase Gunnoe, *Trains* magazine, June 2016, page 24. Also, "Heaviest Train," Andy Cummings and John Godfrey, *Trains* magazine, February 2012, pages 38-49 and "The Long and the Short of Distributed Power," Keith Barrow, *Railway Age*, August 2011, pages 14-18.

172. "LEADER: The Train Handling and Energy Management System" brochure, New York Air Brake website.

173. "Making Railroad History," *Norfolk and Western Magazine*, January 1, 1968. Also see "N&W's Longest Train," Norfolk & Western Historical Society, May 6, 2004, and "Longest and Heaviest Train," *Norfolk and Western Magazine*, November 1967.

174. "Train Records: The Fastest, Longest & Heaviest," www.railserve.com.

175. "CSX Wants to End Locomotive Pusher Operations," Chase Gunnoe, *Trains* magazine newswire, July 21, 2017.

176. For a summary of EPA Tier 0 – 4 emissions standards, please see "Locomotives: Exhaust Emissions Standards." Another excellent resource on these emissions standards is https://www.dieselnet.com/standards/us/loco.php.

177. See EPA July 1, 2012, rule-making for 40 CFR Part 1033, Sub-Part H, "Averaging, Banking, and Trading for Certification," 1033.701 – 1033.750, pages 37230 – 37234.

178. Ibid.

179. "Locomotive – Different Paths, Same Destination," *Trains* magazine, February 2017, page 20.

180. Conversation with Class I railroad manager, September 14, 2017. Also note that a $3 million cost is provided in *Transitioning to a Zero or Near-Zero Emission Line-Haul Freight Rail System in California: Operational and Economic Considerations,* Rail Transportation and Engineering Center (RAILTEC), page xv.

181. "EMD Gets Metrolink Tier 4 Locomotive Order," Douglas John Bowen, *Railway Age*, December 20, 2014; and "Siemens Wins Additional Locomotive Contract in the US," Siemens press release, March 18, 2014.

182. *Sustainable Freight Transport*, California Air Resources Board (CARB), California Environmental Protection Agency; and "CARB Draft Technology Assessment: Freight Locomotives," CARB, April 2016, Table ES-5, page ES-11.

183. GE claims that the avoidance of aftertreatment technology in its ET44AC (Tier 4-compliant) locomotive saved the railroads more than $1.5 billion in urea infrastructure and operational costs. Included in this savings estimate are avoided costs associated with not needing to install a 4,000-pound catalytic converter in every locomotive. According to Mike Iden of Union Pacific Railroad, size considerations of SCR aftertreatment as well as particulate reduction treatment technologies were also concerns for an already crowded locomotive. See Iden's presentation "Locomotive Emissions Technology: Progress and Direction, at the West Coast Diesel Emissions Reduction Collaborative Conference, March 21-22, 2005.

184. "Transitioning to a Zero or Near-Zero Emission Line-Haul Freight Rail System in California: Operational and Economic Considerations," Rail Transportation and Engineering Center (RAILTEC), page 9.

185. "Tier 4 Locomotives Take to the Tracks," David Lustig, *Railway Gazette International*, December 2015, page 30.

186. "Motive Power Review," Chris Guss, *Locomotive 2016*, *Trains* magazine special edition, 2016, page 17.

187. Conversation with Class I railroad locomotive manager, October 2017.

188. See "GE Evolution Series Tier 4 Locomotive, GE Transportation website, and California Dreamin': New Locomotives Take Clean Air Tech from Theory to Reality, Inside Track, November 11, 2016.

189. "Comparative Evaluation of Rail and Truck Fuel Efficiency on Competitive Corridors," page 38. Also note: The Manufacturers of Emissions Controls Association, in its April 2009 "Mobile Source Technologies to Reduce Greenhouse Gas Emissions" background paper states that heavy trucks are 3% to 5% more fuel efficient using diesel particulate filters and SCR systems than filters and ERG systems (page 5).

190. "California Dreamin'. Statement about GE's Tier 4 locomotive on Union Pacific website by Mike Iden, Director of Car and Locomotive Engineering, Union Pacific Railroad.

191. "Tier 4 Locomotives Take to the Tracks," *Railway Gazette International*, page 31.

192. A particulate oxidation catalyst captures particulate emissions without blocking exhaust gas flow. Nitrogen oxide, generated by

the catalyst, reacts with carbon soot to produce carbon dioxide which then exits the exhaust stack. See "Locomotive Builders Continue to Craft Tier 4 Models to Help Railroads Further Their Environmental Pursuits," Jeff Stagl, *Progressive Railroading*, January 2016; and "Particle Oxidation Catalysts," Addy Majewski, Dieselnet Technology Guide.

193. "Locomotive Emissions R&D for the Future: An EMD View," Dave Brann, Locomotive Emissions and System Efficiency Workshop, January 30-31, 2001.

194. "Diesel Exhaust Fluid (DEF) Q & A," Cummins Bulletin, Cummins, 2009.

195. A composite of statements heard by the author from a number of railroad industry locomotive diesel engine experts who have asked to remain anonymous.

196. *Railroad and Locomotive Technology Roadmap*, page 12.

197. "Summary and Analysis of Comments: Control of Emissions of Air Pollution from Locomotive Engines and Marine Compression Ignition Engines Less Than 30 Liters per Cylinder," U.S. Environmental Protection Agency, March 2008, EPA420-R-08-006, pages 11-1 through 11-8.

198. As per a discussion with a mechanical engineer employed by one of the major locomotive manufacturers. Note also that (as previously stated) The Manufacturers of Emissions Controls Association, in its April 2009 "Mobile Source Technologies to Reduce Greenhouse Gas Emissions" background paper states that heavy trucks are 3% to 5% more fuel efficient using diesel particulate filters and SCR systems than filters and ERG systems (page 5).

199. "Executive Order B-32-15," California Governor Edmund G. (Jerry) Brown, Jr., July 17, 2015.

200. *Sustainable Freight Transport*, California Air Resources Board, California Environmental Protection Agency; and *CARB Draft Technology Assessment: Freight Locomotives*, CARB, April 2016.

201. *CARB Draft Technology Assessment: Freight Locomotives*, page V-2 through 3.

202. Ibid., pages V-3 through 7 and V-14 through 15. Additionally, batteries are discussed comprehensively in chapter VI, beginning on page VI-1.

203. Ibid., ES-10 through 12.

204. Ibid., ES-11.

205. "Comments of GE Transportation on California's Draft Sustainable Freight Report," Jennifer Shea, GE Transportation, July 13, 2016.

206. CARB April 13, 2017, petition to the EPA and supportive letters are available on the "Rail Emission Reduction Program" page of CARB's website, https://ww2.arb.ca.gov.

207. "Consensus on Consensus: A Synthesis of Consensus Estimates on Human-Caused Global Warming," John Cook et. al., *Environmental Research Letters*, April 13, 2016.

208. See climate change assessment reports by the Intergovernmental Panel on Climate Change at www.ipcc.ch. Also see "Scientific Consensus: The Earth's Planet Is Warming," Global Climate Change, NASA.

209. "Cost-Benefit Analysis of Alternative Fuels and Motive Designs," Chris Pinney and Brian Smith, Transportation Technology Center, sponsored by the Federal Railroad Administration, U.S. Department of Transportation, April 2013, page 44.

210. Based on conversation with Claudio Filippone, October 9, 2017. ThermaDynamics's website is: www.thermarail.com.

211. The exception to this rule is the use of recovered braking energy by new passenger locomotives to partially meet HEP loads.

212. *Railroad and Locomotive Technology Roadmap*, page 21.

213. "Railpower Technologies Corp. Green Goat Hybrid Locomotive Roster," Jody Moore, March 12, 2008.

214. RailPower Locomotive section, R.J. Corman's website does not show the Green Goat for sale.

215. "Green Goat Finally Goes Big-Time," David Lustig, *Trains* magazine, June 2005, page 20.

216. *Comparative Evaluation of Rail and Truck Fuel Efficiency on Competitive Corridors*, page 22.

217. See "GE Introduce Hybrid Locomotive to Public," Green Car Congress, May 27, 2007; and "GE Unveils Hybrid Locomotive," *Railway Gazette*, July 2, 2007.

218. Ibid; Also, "Rechargeable Batteries," R. Edwin Garcia, University of Purdue, slide 3; and "GE Transportation's Evolution Hybrid Locomotive Makes Chicago Debut at Railway Technology Exhibit," *Business Wire*, September 12, 2008.

219. "Erie-Built Hybrid Locomotive Ready When Buyers Are," Jim Martin, *Erie Times-News*, June 20, 2010.

220. Ibid.

221. See Carbon Tax Center and Citizen's Climate Lobby for discussions of carbon tax and carbon fee. The Citizen's Climate Lobby advocates a revenue-neutral fee that rebates to American households all funds collected by a GHG emissions-based fee placed on fossil fuels as they enter the economy. The CCL calls for an initial $15/ton fee to increase by $10/ton/year until emissions are 10% of 1990 levels. The politics of deliberately increasing the price of fossil fuels is difficult, to say the least, even if essential to reduce fossil fuel use and GHG emissions.

222. "The Imperative of a Carbon Fee and Dividend," James E. Hansen, Climate Science, Awareness and Solutions blog, Earth Institute, Columbia University.

223. *CARB Draft Technology Assessment: Freight Locomotive*, page VI-6.

224. Ibid., VI-9.

225. "Comments of GE Transportation on California Air Resources Board's Draft Technology Assessment for Freight Locomotives," Jennifer Shea.

226. GB Rail Powertrain Efficiency Improvements, pages 118 and 140.

227. Ibid., page 70.

228. *Railroad and Locomotive Technology Roadmap*, page 23.

229. "GE Introduce Hybrid Locomotive to Public." Also see "21st Century Locomotive Technology (Locomotive Systems Tasks)," Robert D. King, Lembit Salasoo, and Paul Houpt, GE Global Research, April 2006.

230. "21st Century Locomotive Technology: Quarterly Technical Status Report 28 DOE/AL68284-TSR28," page 1, U.S. DOE Office of Scientific and Technical Information (OSTI), www.osti.gov.

231. "GE Plant to Commercialize Hybrid Loco Batteries," *Railway Gazette*, May 13, 2009; "GE Charges Ahead with New Battery Plant," Yi-Ke Peng, *Albany Times Union*, September 29, 2011; GE Proclaims Success Despite Battery Plant Closure," Lauren Stanforth, *Albany Times Union*, January 12, 2016.

232. See presentation by Ed Hall, General Manager of GE Transportation, at the "Transitioning to Zero Emission Freight Transport Technologies Symposium," April 11, 2013, sponsored by California South Coast Air Quality Management District (SCAQMD). Summary slide that states "Near term development of battery locomotive is not possible – Engineering Resources focused on Tier 3/Tier 4/LNG; Battery commercialization not complete."

233. "EcoActive Technologies: MITRAC Hybrid–The Dual Power Propulsion Chain," www.bombardier.com.

234. For a good discussion of the pros and cons of some of these energy storage media for use in locomotives, see "Future Power Technologies - Final Report," Mohammad Rasul, Yan Sun, and Mitchell McChanachan, CRC for Rail Innovation, Australian Government Initiative, 2013, pages 9-15.

235. See "Hybrid Energy Storage Systems," Rail Propulsion Systems website and "Future Power Technologies - Final Report," Mohammad Rasul, Yan Sun, and Mitchell McChanachan, page 9.

236. "A Battery Powered Alternative - NS 999 Generation 2.0 – The Next Step in Battery Powered Locomotives," Norfolk Southern *2014 Sustainability Report*.

237. Ibid.

238. TransPower, www.transpowerusa.com.

239. *CARB Draft Technology Assessment: Freight Locomotive*, pages VI-10 through 11.

240. Calculated at the lower heating value of diesel fuel, i.e., 128,000 BTUs per gallon.

241. "Zero Emission Booster Locomotive" is California Air Resources Board terminology that describes a locomotive as "zero emissions" if it emits zero criteria air pollutants (NOx, HC, PM, and CO) at point of use.

242. "On Board Electrification and Near-Zero Emissions for Regional Rail," Dave Cook and Ian Stewart, September 2014, and "Concept Definition of a Zero Emissions Booster Locomotives for Regional Passenger Rail," Dave Cook, Peter Eggleton, and Ian Stewart, 2015 Joint Rail Conference, ASME paper JRC2015-5743.

243. "Primove – true e-mobility," www.primove.com. Also: "Suppliers Eye Market for 'Hybrid' Streetcars," Douglas John Bowen, *Railway Age*, August 2011.

244. *Transitioning to a Zero or Near-Zero Emission Line-Haul Freight Rail System in California: Operational and Economic Considerations*, Rail Transportation and Engineering Center (RAILTEC), page xv.

245. For a Union Pacific perspective see "Liquefied Natural Gas (LNG) as a Freight Railroad Fuel: Perspective from a Western U.S. Railroad," Mike Iden, presentation at ASME 2012 Rail Transportation Division Fall Technical Conference, October 2012.

246. "Getting All Gassed Up," *Trains* magazine, November 2016, pages 18-19.

247. "Florida East Coast Puts LNG Locomotives in Revenue Service," *Trains* magazine newswire, June 15, 2016.

248. CARB estimates the cost of the LNG locomotive retrofit kit to be $500,000; and "Cooking with Gas," William Vantuono, *Railway Age*, December 2017, pages 20 -23.

249. "For Progress Rail/EMD, Two LNG Systems," William Vantuono, *Railway Age*, September 20, 2016.

250. *Transitioning to a Zero or Near-Zero Emission Line-Haul Freight Rail System in California: Operational and Economic Considerations*, Rail Transportation and Engineering Center (RAILTEC), page xiv. Also see "Dual Fuel Locomotive Is Friendly to the Environment," GE Global Research webpage, 2017.

251. *Transitioning to a Zero or Near-Zero Emission Line-Haul Freight Rail System in California: Operational and Economic Considerations*, Rail Transportation and Engineering Center (RAILTEC), page 11.

252. See slide 5 in "Fuels of the Future for Cars and Trucks," Dr. James J. Eberhardt, 2002 Diesel Engine Emissions Reduction (DEER) Workshop, San Diego, California, August 25–29, 2002. Also, note that LNG/CNG fuel tanks would require even more space because they are high pressure and therefore of cylindrical shape. Plus, LNG tanks must have an insulating vacuum space around them and CNG tanks require thick walls because of their extremely high pressure.

253. CARB estimates the cost of the LNG tender (serving two locomotives) to be $1 million. See *CARB Draft Technology Assessment: Freight Locomotive*, page IV-11.

254. "Experts Weigh In on LNG," William C. Vantuono, *Railway Age*, September 6, 2013.

255. *CARB Draft Technology Assessment: Freight Locomotive*, page IV-12.

256. A "diesel gallon equivalent" (DGE) is the amount of natural gas that has the same number of BTUs as a gallon of diesel fuel. A gallon of diesel fuel has a BTU content of 137,000 BTUs while an MCF (1,000 cubic feet) of natural gas has a BTU content of 1,020,000 BTUs. Thus, a DGE of natural gas is 137,000/1,020,000 = 0.134 MCF. The $1.50 price differential required for a 5-year payback would be achieved when, for example, diesel fuel costs $3.00/gallon and .134 MCF of natural gas costs $1.50 less or $1.50. At that rate, one MCF would cost $1.50/.134 = $11.19.

257. *Transitioning to a Zero or Near-Zero Emission Line-Haul Freight Rail System in California: Operational and Economic Considerations*, Rail Transportation and Engineering Center (RAILTEC), page 12.

258. "Understanding Greenhouse Gas Emissions," U.S. Environmental Protection Agency website.

259. "Methane Leaks from North American Natural Gas Systems," A.R. Brandt et al., *Science* magazine, February 14, 2014. pages 733-735.

260. Conversation with Environmental Defense Fund scientist on August 25, 2017, plus see: "Methane: the Other Important Greenhouse Gas," and "Synthesis of Recent Ground-Level Methane Emission Measurements from the U.S. Natural Gas Supply Chain," James Littlefield et al., *Journal of Cleaner Production*, January 2017. Note that top-down studies generally show higher levels of methane leakage than ground-level studies.

261. See "Methane Emissions and Climatic Warming Risk from Hydraulic Fracturing and Shale Gas Development: Implications for Policy," Robert Howarth, *Energy and Emissions Control Technologies*, October 8, 2015.

262. "Greenhouse Gas and Noxious Emissions from Dual Fuel Diesel and Natural Gas Heavy Goods Vehicles, Marc E.J. Stettler, *Environmental Science & Technology*, American Chemical Society, January 12, 2016.

263. As per conversations with manufacturer and railroad locomotive experts. Also see, "Pumps-to-Wheels Methane Emission from Heavy-Duty Transportation Sector," Nigel N. Clark et al., *Environmental Science & Technology*, American Chemical Society, December 22, 2016. Clark and his colleagues estimate downstream methane emissions for heavy-duty (non-railroad) vehicles to be 13.1 grams per kilogram of natural gas fuel, i.e., 1.3%; and "Greenhouse Gas and Noxious Emissions from Dual Fuel Diesel and Natural Gas Heavy Goods Vehicles," Marc E. J. Stettler, *Environmental Science & Technology*, American Chemical Society, January 12, 2016, which highlights the need for in-cylinder strategies to reduce methane emissions.

264. Ibid.

265. "NRDC Policy Basics: Hydrofracking," Natural Resources Defense Council website, February 2013.

266. "Thanks to Fracking, Earthquake Hazards in Oklahoma Now Comparable to Parts of California," James Conca, *Forbes* magazine, September 7, 2016.

267. "Freight Railroad Energy: Alternatives & Challenges," Michael E. Iden, presentation at RAILTEC conference, University of Illinois at Urbana-Champaign, February 15, 2013, slide 14.

268. "Emissions from Trucks Using Fischer-Tropsch Diesel Fuel," Paul Norton et al., U.S. Department of Energy, Alternative Fuels Data Center, 1998.

269. *Tackling Climate Change Through Livestock: A Global Assessment of Emissions and Mitigation Opportunities*, Food and Agriculture Organization of the United Nations, 2013.

270. "Locomotive Emissions Measurements for Various Blends of Biodiesel Fuel," Steven Fritz, John Hedrick, and James Rutherford, SAE International Technical Paper #2013-24-0106, September 8, 2013.

271. "Locomotive Biofuel Study: Preliminary Study of the Use and Effects of Biodiesels in Locomotives," Melissa Shurland (FRA), Wade Smith (Amtrak), and Steve G. Fritz (SwRI), Federal Railroad Administration, U.S. Department of Transportation, May 2014.

272. "New Additive Hands Biodiesel the Win as Cleanest Liquid Fuel in the U.S.," National Biodiesel Board press release, July 27, 2017. And "CARB Executive Order G-714-ADF01 – Certification of Alternative Diesel Fuel Resulting in Emissions Equivalence with Diesel – National Biodiesel Board - VESTA™1000," July 20, 2017.

273. "Regulation of Fuels and Fuel Additives: Changes to Renewable Fuel Standard Program – Final Rule," (EPA-HQ-OAR-2005-0161; FRL-9112-3), 40 CFR Part 80, U.S. Environmental Protection Agency, *Federal Register*, March 26, 2010, pages 14,788-14,790.

274. "Reassessment of Lifecycle Greenhouse Gas Emissions for Soybean Oil Production," A. Pradhan et al., 2012 *American Society of Agricultural and Biological Engineers*, Volume 55(6): 2257-2264.

275. "Everything You Ever Wanted to Know about Biodiesel," Jeremy Martin, Union of Concerned Scientists, June 22, 2016.

276. Ibid.

277. "Renewable Fuel Standard Program: Standards for 2017 and Biomass Based Diesel Volume for 2018," 40 CFR Part 80, U.S. Environmental Protection Agency (EPA-HQ-OAR-2016-0004; FRL-9955-84-OAR), *Federal Register*, December 12, 2016.

278. The Locomotive Maintenance Officers Association (LMOA) is a group of international railroad officers whose responsibilities and interests are concerned with the repair and maintenance of railroad locomotives and shop-related equipment.

279. "Multimedia Evaluation of Renewable Diesel," Multimedia Working Group, CARB, May 2015, Biodiesel and Other Renewable Diesel Fuels, National Renewable Energy Laboratory, U.S. Department of Energy; and "Renewable Diesel Fuel," Robert McCormick and Teresa Alleman, National Renewable Energy Laboratory, July 18, 2016.

280. "LCFS Pathway Certified Carbon Intensities," CARB, October 13, 2017, summarizes CARB's latest analysis of the carbon intensities of alternative fuels compared to diesel fuel.

281. "Renewable Diesel Use in California Moves to Fast Track," Isha Salian, *San Francisco Chronicle*, September 14, 2017.

282. "Carbon Intensity Lookup Table for Gasoline and Fuels that Substitute for Gasoline, CARB, 2012.

283. "Final Draft Report - Feasibility of Renewable Natural Gas as a Large-Scale, Low Carbon Substitute," Amy Myers Jaffe, Institute of Transportation Programs, UC Davis, prepared for CARB, 2017; and "Study Finds Renewable Natural Gas Could Meet ~85% of Current Natural Gas Use in Transport in California by 2020; Much Higher Volumes Possible with Right Policies," Green Car Congress, December 5, 2016.

284. 90.6 billion cubic feet of methane has the energy content of 656 million gallons of diesel fuel. This is equal to 3 million Notch 8 hours of locomotive operation at 210 gallons/hour.

285. "Hydrogen Production Processes," Office of Energy Efficiency and Renewable Energy, U.S. Department of Energy.

286. "Fuel Cell Prototype Locomotive," David Barnes, Project Director; Vehicles Project, LLC; U.S. Department of Energy (Grant Award No. DE-FG36-05GO85049), June 2007, page 3. Also see, "Demonstration of a Hydrogen Fuel Cell Locomotive," Kris S. Hess et al., American Public Transportation Association, 2010 Technical Forum.

287. Hydrogen can be produced from natural gas (methane) by subjecting it to very high temperature steam. The process is called natural gas reforming. See: "Hydrogen Production: Natural Gas Reforming," Office of Energy Efficiency and Renewable Energy, U.S. Department of Energy.

288. "Hydrogen Production and Distribution," Alternative Fuels Data Center, U.S. Department of Energy website.

289. *CARB Draft Technology Assessment: Freight Locomotive*, pages VII-11 and VII-7 through VII-13.

290. Ibid., pages VII-2 through VII-7.

291. Ibid., page VII-6.

292. Ibid., page VII-13.

293. Ibid., page VII-12.

294. "Coradia iLint," Alstom Partners, www.partners.alstom.com; and "iLint: the World's First Hydrogen-Powered Train," www.Railway-Technology.com, January 22, 2018.

295. "Ontario on Track for Next Generation of Clean Trains?," Ben Spurr, *Toronto Star*, November 10, 2017.

296. "Regional Express Rail Program Hydrail Feasibility Report," prepared for Metrolinx by H2M HILL Canada Limited (now Jacobs Engineering Group Inc.), Ernst & Young Orenda Corporate Finance Inc., and Canadian Nuclear Laboratories, February 2, 2018; "Ontario Taking Next Steps in Testing Hydrogen-Powered Train Technology," Ontario Ministry of Transportation, February 22, 2018; "Ontario Hosts International Symposium on Hydrogen-Powered Rail," Ontario Ministry of Transportation, November 16, 2017; and "Electrification," Go Transit website.

297. "Regional Express Rail Program Hydrail Feasibility Report," page 91.

298. "Austria's Zillertal Railway Opts for Hydrogen Trains," Erwin Reidinger, Railjournal.com, February 2, 2018.

299. The ACS-64 is also called the *Amtrak Cities Sprinter* and its short-time rating is 6,400 kilowatts or 8,600 hp. It's continuous rating is equal to 6,700 hp.

300. "Inside Amtrak's New Fleet of High-Tech Train Engines," Sophie Reid, *The Pulse* (WHYY-FM), May 15, 2014.

301. "Amtrak ACS-64: Speed, Power, Efficiency," Michael Latour, *Railway Age*, May 31, 2011.

302. "Cutting Traction Power Costs with Wayside Energy Storage Systems in Rail Transit Systems," L. Romo, D. Turner, and L.S.B. Ng, ASME/IEEE Joint Rail Conference, Pueblo, CO, 2005, pages 187-192.

303. "Inside Amtrak's New Fleet of High-Tech Train Engines."

304. Conversation with Amtrak locomotive manager, July 13, 2017; also *Railroad and Locomotive Technology Roadmap*, page 21 (where 64% is given).

305. Locomotive component efficiencies and overall efficiency informally confirmed with Siemens corporation, October 16, 2017.

306. Conversation with electric locomotive manager, August 31, 2017, though this number (low 90% for electric locomotive efficiency) may not have included inefficiencies associated with locomotive gearing and journal friction.

307. "Breaking the Power Plant Efficiency Record," GE Power, www.gepower.com.

308. "How Much Electricity Is Lost in Transmission and Distribution in the United States?", Frequently Asked Questions, U.S. Energy Information Administration. Five percent average loss is given.

309. *When the Steam Railroads Electrified*, William D. Middleton, Milwaukee, WI (Kalmbach Publishing), 1974 and 1976.

310. "Electrification: Is It Going to Happen?," William D. Middleton, *Railway Age*, March 10, 1975, pages 28-37.

311. "Electrification: Who Will Pay?," Richard Fishbein, *Railway Age*, July 11, 1977, pages 22-26.

312. An exception to the rule that U.S. freight railroads are not electrified is the "Black Mesa and Lake Powell Railroad." This isolated electric railroad uses 1970s-vintage GE E60 electric locomotives to haul coal from a Peabody coal mine near Kayenta, AZ, to the Navajo Generate Station in Page, AZ. Approximately 3 roundtrip trips per day are made over the 78-mile route.

313. "Innovative Traction and Electrification in Bayern Plan to Reduce Diesel Dependency," *Railway Gazette*, January 30, 2018.

314. "Italian High-Speed Rail Freight Plan Unveiled," Marco Chiadoni, *International Railway Journal*, April 10, 2018.

315. "Road mileage" here means route-miles, not counting parallel tracks, yards and sidings.

316. "Class I Railroad Statistics," Association of American Railroads, May 1, 2017.

317. *Solutionary Rail* website.

318. Rail Solution and Steel Interstate Coalition websites.

319. "The Need for Freight Rail Electrification in Southern California," Brian Yanity, Californians for Electric Rail, September 17, 2017.

320. Amtrak purchased 70 ACS-64 Siemens electric locomotives for $466 million in 2010 that equals $6.7 million per locomotive – thus the $5 million per locomotive cost used here may be an underestimate, though the Amtrak locomotives were equipped with head-end power that would not be needed for freight locomotives.

321. Not surprising, there are different estimates for the cost to electrify: (1) Ontario Metrolinx's Go Transit 2010 electrification study projects an electrification infrastructure cost of C$2.369 billion for the entire 316-mile system Go Transit rail system. While a double- and single-track breakdown was not available, that works out to $7.1 million/mile, after adjusting for the currency difference at this writing. (2) "Transitioning to a Zero or Near-Zero Emission Line-Haul Freight Rail System in California: Operational and Economic Considerations," Rail Transportation and Engineering Center (RAILTEC), provides $4.8 million/mile (2012) on page 20. (3) "Reducing Emissions in the Rail Sector: Technology and Infrastructure Scan and Analysis," R. Barton and T. McWha, National Research Council Canada, September 2012, provides a $950,000 - $1,120,000 per kilometer figure in pre-2009 dollars on page 15. Adjusting to miles and 2017 dollars, those costs would be $1.74 million - $2.06 million/mile, or, say, $1.9 million/mile. And (4) an Amtrak manager told the author in July 2017 that the cost of electrification infrastructure was $1.1 million per mile for two track installations plus $11 million every 10 miles for substations. That works out to $2.2 million/mile, including catenary wire, pole, and substations only. This number would have to be increased 10% for inflation, $2.2 million x 1.1 = $2.42 million. This cost estimate was for a 50 KW distribution system but could increase substantially depending on overhead obstacles encountered en route. If we take an average of these four estimates, we get ($7.1M + $4.8M + $1.9M + $2.4M)/4 = $4.04 million per mile.

322. Catenary costs of $12-24 million per mile are given in "Moving California Forward: Zero and Low-Emission Goods Movement Pathways," A Report to the California Cleaner Freight Coalition, Gladstein, Neandross and Associates, November 2013, pages 34-35.

323. "Transitioning to a Zero or Near-Zero Emission Line-Haul Freight Rail System in California: Operational and Economic Considerations," Rail Transportation and Engineering Center (RAILTEC), page 63.

324. The terms "carbon tax" and "carbon fee" have been used interchangeably here but proponents define them differently. The carbon tax is just that, a tax that generates revenue for government spending (presumably on clean energy), and the carbon fee would be revenue neutral and returned to Americans via periodic checks to them from the U.S. Treasury. Both the carbon tax and carbon fee would make the price of fossil fuel rise at a steady rate. Both would collect funds from fossil fuel companies for fuel as it enters the economy based on carbon content or relative GHG emissions per unit of energy.

325. For explanation of the dual-mode locomotive technology see "EcoActive Technologies: MITRAC Hybrid–The Dual Power Propulsion Chain," www.bombardier.com.

326. "NJT's ALP-45DP Enters Revenue Service," William C. Vantuono, Railway Age, May 30, 2012.

327. Railroad Facts 2016, Association of American Railroads, 2016, page 63.

328. See "How Much Electricity Is Lost in Transmission and Distribution in the United States," Frequently Asked Questions, U.S. Energy Information Administration.

329. Discussion with Amtrak manager, August 31, 2017.

330. "High Speed Euro Train Gets Green Boost from Two Miles of Solar Panels," The Guardian, June 6, 2011.

331. "World's Largest Solar Bridge Opens," The Guardian, January 22, 2014.

332. "Renewable Energy, Additionality, and Impact: An FAQ on the U.S. Voluntary Renewable Energy Markets," Timothy Juliani, Edison Energy white paper, January 19, 2018.

333. "Via Rail Seeks Private Sector Partnership for Toronto-Montreal Upgrade," David Thomas, International Railway Journal, August 17, 2016.

334. "National Dream Redux," Railway Age, April 2016, page 22.

335. ""Power Struggle," Sam Howe Verhovek, New York Times Magazine, January 12, 1992, and "Canadian Hydro Project Opposed," Catherine Foster, Christian Science Monitor, March 21, 1991.

336. "Climate Protection at Deutsche Bahn," Deutsche Bahn website, July 2017. Also see "Trains That Run Like, and on, the Wind," Erik Kirschbaum, New York Times, August 21, 2011.

337. "James Hansen on Nuclear Power," video, American Nuclear Society website; "Climate Change Warriors: It's Time to Go Nuclear," Tom Patterson, CNN, November 3, 2013; "Nuclear Power Paves the Only Viable Path Forward on Climate Change," James Hansen, Kerry Emanuel, Ken Caldeira, and Tom Wigley, The Guardian (US Edition), December 3, 2015; "Advancing Nuclear Energy to Help Address Climate Change and Air Pollution," James Hansen, December 29, 2015; Storms of My Grandchildren, James Hansen, 2009, pages 194-204. Also see "Pandora's Promise" film directed by Roger Stone, 2013.

338. Storms of My Grandchildren, pages 200-204.

339. "What Is U.S. Electricity Generation by Energy Source?," Frequently Asked Questions, U.S. Energy Information Agency, July 2017 (2016 data).

340. Ibid.

341. "A Technology Roadmap for Gen IV Nuclear Energy Systems," U.S. DOE Nuclear Energy Research Advisory Committee and the Generation IV International Forum, December 2002.

342. This refers to the efficiency of converting the energy in nuclear fuel into heat, not the efficiency of converting heat produced by that fuel into electricity. The latter still might hover around 33%.

343. "Nuclear Power: Climate Fix or Folly," Amory Lovins, Imran Sheikh, and Alex Markevich, Rocky Mountain Institute, December 2008; "Nuclear Power: Economics and Climate-Protection Potential," Amory Lovins, Rocky Mountain Institute, January 2006; and other articles found on the Rocky Mountain Institute "Nuclear Energy" search page.

344. "Why Jim Hansen Is Wrong About Nuclear Power," Joe Romm, Climate Progress, January 7, 2016.

345. "Does the World Need Nuclear Energy?," TED Talk debate – Stewart Brand vs. Mark Jacobson, February 2010; "Mark Jacobson to James Hansen: Nukes Are Not Needed to Solve the World's Climate Crisis," Mark Jacobson, EcoWatch website, January 4, 2016. Also, "The United States Can Keep the Grid Stable at Low Cost with 100% Clean, Renewable Energy in All Sectors Despite Inaccurate Claims," Mark Z. Jacobson et al., PNAS, June 27, 2017.

346. "The Murky Future of Nuclear Power in the United States," Diane Cardwell, *New York Times*, February 18, 2017. Also see "Nuclear Power Plants, Despite Safety Concerns, Gain Support as Energy Sources," Diane Cardwell, *New York Times*, May 31, 2016.

347. "Cyberattacks Put Russian Fingers on the Switch at Power Plants, U.S. Says," Nicole Perlroth and David E. Sanger, *New York Times*, March 15, 2018; and "Russian Government Cyber Activity Targeting Energy and Other Critical Infrastructure Sectors," United States Computer Emergency Readiness Team, Department of Homeland Security, March 15, 2018.

348. "A Complete Guide to Carbon Offsetting," Duncan Clark, *The Guardian*, September 16, 2011.

349. "Should You Buy Carbon Credits? – A Practical and Philosophical Guide to Neutralizing Your Carbon Footprint," Natural Resources Defense Council, www.nrdc.org.

350. Norfolk Southern 2016 Sustainability Report, pages 5 and 70.

351. Reforestation is a challenging mechanism for creating carbon credits because (a) forests are not permanent, (b) their rate of carbon sequestration is modest, and (c) sequestration is affected by many factors including tree type and age. Consider Norfolk Southern's Trees and Trains program will plant 6.04 million native hardwood and soft woods trees on 10,000 acres (roughly 20 square miles). If we assume that an afforestation rate for land that was previously cropland/pasture of 6 tons CO_2e per acre, then these trees will sequester 6 x 10,000 = 60,000 tons CO_2e. If 22.4 pounds of CO_2 are emitted by burning a gallon of diesel fuel and a well-utilized locomotive can consume 300,000 gallons of diesel fuel per year, then that locomotive will emit 6,720,000 pounds or 3,360 tons of CO_2 per year. At that rate, the 10,000 acres of planted trees will sequester the emissions of only 18 locomotives, underlining the importance of energy conservation as a primary strategy for avoiding carbon emission in the first place. (See Norfolk Southern 2016 Sustainability Report, page 58; and "U.S. Tree Planting for Carbon Sequestration," Ross W. Gorte, Congressional Research Service, May 4, 2009, where the sequestration rate for reforesting cropland/pasture was given as 2.2 to 9.5 tons of CO_2e.)

352. "Power Forward 3.0: How the Largest Companies Are Capturing Business Value While Addressing Climate Change," World Wildlife Fund, Ceres, Calvert, and CDP (formerly the Carbon Disclosure Project), 2017, page 2.

353. Amtrak 2015 Sustainability Report, page 44; "Amtrak Publishes 2016-2017 Sustainability Report," March 19, 2018.

354. BNSF 2015 Sustainability Report, pages 20 and 13.

355. Canadian National "Environment" webpage.

356. Canadian Pacific 2014 Sustainability Report, page 23.

357. CSX 2016 Sustainability Report, page 21, and CSX 2015 Sustainability Report, pages 48 and 53.

358. *Kansas City Southern 2015 Sustainability Report*, page 46.

359. *Norfolk Southern 2016 Sustainability Report*, page 65.

360. *Union Pacific 2015 Sustainability Report*, pages 37 and 36.

361. Ibid., page 38. If Union Pacific sold these locomotives to other railroads and they are in use, net overall efficiency and emissions gain would be reduced. However, these locomotives may be replacing even less efficient, more polluting locomotives at these other railroads.

362. *BNSF 2015 Sustainability Report*, page 14.

363. "Would Saving a Livable Climate Destroy Buffett's Fossil Fuel Empire?," Joe Romm, ThinkProgress, www.thinkprogress.org.

364. "A Letter from the CEO," GE sustainability website.

365. *Caterpillar Building Better 2016 Sustainability Report*, Caterpillar website, www.caterpillar.com.

366. "Siemens Is Going Carbon Neutral," Siemens.com website.

367. "Coal – A Twisted Future," Peter A. Hansen, *Trains* magazine, March 2016, pages 40-47; Also "Coal – The Lifeblood of American Railroading," *Trains* magazine, April 2010, pages 26-49; and "Heavy Hauls," *Trains* magazine special edition, 2015, for an updated and abbreviated version of the 2010 article.

368. "Railroads and Coal," Association of American Railroads, July 2017, page 1.

369. "Short-Term Energy Outlook," U.S. Energy Information Administration website.

370. "Before the Fall – Could We Have Foreseen the Decline in Freight Traffic? Or Avoided It?," Michael W. Blaszak, *Trains* magazine, March 2017, page 29.

371. "Railroads and Coal," page 6.

372. "Can Policy and Appointees Save King Coal," David Nahass, *Railway Age*, January 2017, page 64; "Coal – A Twisted Future," Peter A. Hansen, *Trains* magazine, March 2017, pages 40-47; "Rethinking Coal in Current Context," Tony Hatch, www.progressiverail.com, August 2017; "Trump's Policies Aid Freight Railroads: Early Trend Sees Coal as Biggest Beneficiary," Don Phillips, *Trains* magazine, December 2017; and "Coal Bump in 2017 Not Long Slump's End," Bill Stephens, *Trains* magazine, December 2017.

373. "Coal – A Twisted Future," page 47.

374. "Wind Power Shipments on the BNSF Network," BNSF website.

Chicago, IL, November 2015. Photo Credit: Judie Simpson.

Index

Note: Locomotive types are listed by manufacturer. EMD locomotives (General Motors and Progress Rail EMD) are all listed under Progress Rail/EMD Locomotives.

Locomotive diesel engine (prime mover) technologies are listed separately and also under diesel engine.